The Self-organizing University

Alan Bain · Lucia Zundans-Fraser

The Self-organizing University

Designing the Higher Education Organization for Quality Learning and Teaching

Springer

Alan Bain
Faculty of Arts and Education
Charles Sturt University
Bathurst, NSW
Australia

Lucia Zundans-Fraser
Faculty of Arts and Education
Charles Sturt University
Bathurst, NSW
Australia

ISBN 978-981-10-4916-3 ISBN 978-981-10-4917-0 (eBook)
DOI 10.1007/978-981-10-4917-0

Library of Congress Control Number: 2017940311

Printed on acid-free paper

This Springer imprint is published by Springer Nature
The registered company is Springer Nature Singapore Pte Ltd.
The registered company address is: 152 Beach Road, #21-01/04 Gateway East, Singapore 189721, Singapore

We dedicate this book to our families, and to those who think better learning and teaching should be available in all places of learning.

Acknowledgements

This book has benefited from the work of the many colleagues with whom we have worked in universities and schools. We are especially grateful for the contribution of Nicholas Drengenberg and Catherine Newell, who provided extensive feedback on the manuscript and who recognize the challenge and opportunity afforded by better learning and teaching in higher education. We would also like to thank the Inclusive Education Team at Charles Sturt University, Julie Lancaster, Sarah McDonagh, and Greg Auhl whose research, course design, and implementation provided an empirical base for much of what is written here.

Contents

List of Figures

Chapter 1
The Quality Illusion in Learning and Teaching

1.1 Introduction and Overview

This book is about improving the quality of learning and teaching at scale in higher education. All universities have great teachers and groups of academics that enthusiastically pursue curricular innovation. Much more elusive however is the way universities scale up the learning and teaching excellence of individuals and groups to exert a whole-of-organization impact on the quality of the learning experience for all students. While such an impact is frequently claimed in mission statements and marketing materials, we will show what it really takes to substantiate such claims with the models, systems, methods, and tools required to attribute a university's learning and teaching efforts to student learning at scale.

In this chapter, we examine a chain of initiatives commonly employed to establish quality in learning and teaching in universities and identify a problem that undermines them all. We then describe what is required to address the problem by developing a deep understanding of the context for learning and teaching. We introduce the Self-Organizing University (SOU), a model for understanding context and creating systemic change in learning and teaching. To conclude the chapter, we describe the six principles of the SOU and the ways those principles are taken up in subsequent chapters as a guide for creating a SOU.

1.2 Quality in Learning and Teaching—Current Practice

Learning and teaching represents the largest component of a university's operating budget yet we know little about how well this money is spent and the quality of what is produced (Norton and Cherastidtham 2015). Massy et al. (2012) describe the measurement of teaching quality as the elephant in the room in determining the productivity of learning and teaching in higher education noting that none of the

A. Bain and L. Zundans-Fraser, *The Self-organizing University*,
DOI 10.1007/978-981-10-4917-0_1

approaches currently employed are sufficiently robust to be included in a determination of quality. The authors state that:

> …while current and prospective learning and engagement measures are useful in particular contexts, they cannot be brought together into comprehensive, robust, indices for quality adjustment. (p. 6)

According to Sullivan et al. (2012) in the *National Research Council Report, Improving Measurement of Productivity in Higher Education* quality is usually established through price competitiveness or performance studies that show how a particular product or service has better performance characteristics than others. To discern quality, we need to see performance in sustained practice and under conditions where there is confidence about the way it is measured. For example, the crash testing of cars yields credible evidence of their safety that in turn translates into distinctions among vehicle brands that along with other features drive a determination of cost and quality. We expect those features to be present at scale meaning in all iterations of the car not just in the vehicle submitted for testing. In professional service fields like medicine and law, patients may select a surgeon who has a record of successful performance with a particular surgical procedure. Those charged with crimes seek lawyers with successful track records for defending clients. Performance differences translate into price differentiation for the services provided. Variability in performance is expected although the variability exists within a context where all professionals in the field use accepted protocols for professional practice.

In higher education, price competitiveness is confounded by subsidies, government funding, and aid programs that insulate universities from genuine price competition (Massy et al. 2012), while as we will see in this chapter rigorous performance studies are precluded by the current ambiguous circumstances of learning and teaching in higher education. The inability to conduct robust performance studies may seem surprising given the extensive efforts to assure quality and regulate the sector with standards of accreditation, national, state- and university-level initiatives to support learning and teaching; the evaluation of teaching practice within universities; setting frameworks for promotion, and measuring productivity. Figure 1.1 depicts an effort chain of major initiatives employed to address the quality of learning and teaching in universities including those identified by Probert (2015).

The six elements in Fig. 1.1 are intended to provide universities with a deeper understanding of the learning process, and exert an influence on teaching quality and student outcomes at scale. There is also recognition within the sector that to improve the quality of learning and teaching, the discrete elements in the effort chain must be integrated and addressed holistically (Probert 2015). However, for reasons we will explain in this chapter and throughout the book, a scaled, integrated, and holistic approach has not occurred in higher education. We will show that each link in the chain falters because of a failure to fully understand the context in which learning and teaching occurs. Whether referring to learning and teaching

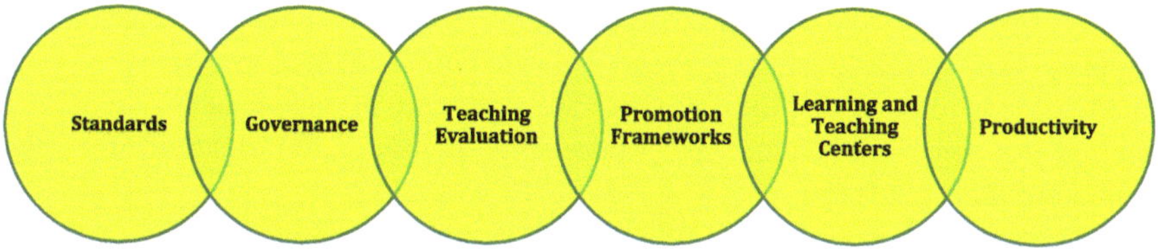

Fig. 1.1 The effort chain for assuring the quality of learning and teaching in Universities (We have connected the initiatives in this figure because in the normal practice of learning and teaching in universities, they are expected to be interrelated although we will show that the coupling is loose at best.)

in design, enactment, and student engagement, a context refers to the way normal work is transacted to produce an expected or intended outcome (Bain and Drengenberg 2016).

1.3 The Problem

The cornerstone of a professional context in any field is comparable and visible professionally controlled practice (Bowker and Star 2000). *Comparability* means the use of evidence-based approaches across multiple settings and individuals. *Visibility* means work processes that are observable and make use of evidence-based approaches in ways that can differentiate among more or less effective practice (Bowker and Star 2000). Comparability and visibility produce *workable distinctions* in routine practice (Drengenberg and Bain 2016). In higher education, workable distinctions mean a visible and comparable difference in the quality of learning and teaching practice in normal work that can be used to make a valid and reliable determination of quality (Drengenberg and Bain 2016). *Professional Control* is a managed professional process that makes standardization and differentiation of practice possible while retaining a degree of professional flexibility and autonomy (Bowker and Star 2000) . The term control is used here to denote a process of dynamic management and evaluation of comparable and visible practice, as opposed to something that restricts or constrains.

All professional fields are built upon an understanding of context derived from the use of research-based protocols for professionally controlled practice (Drengenberg and Bain 2016). A protocol is defined as the structured, evidence-based rules and procedures associated with a visible and comparable professional process. It is not a checklist or an algorithm but a rigorous and expansive foundation for practice. Doctors and nurses employ protocols for surgery, the administration of medication and postoperative care each based on empirical research. Engineers employ design protocols to create the specifications for structures or designing vehicles. Lawyers build their cases on protocols and standards for evidence and case building. All practitioners in their respective fields employ these protocols at scale. They anchor the combination of affective and cognitive factors that mediate professional behavior and account for variance in quality. Most importantly they possess a foundation of visible and comparable practice for problem-solving, for addressing failures, and developing better approaches into the future.

We recognize that in the field of education more generally there is an extensive and contested range of perspectives on the professional standing of the field including some that position education as an exceptional case with respect to professionally controlled practice and the developmental course of professions. It is not our intention to visit this debate here suffice to say that we see the *exceptionalist* arguments about education to be fully consistent with those raised in the history of other fields as they progressed toward genuine professionalism (e.g., McCloskey and Bulechek 1994). Further, when we use examples of other fields we recognize the many miscues, incomplete technologies, and failures that characterize all fields in their developmental course as they pursue quality practice.

In learning and teaching in higher education, an understanding of context would mean that a university could deliver a visible and comparable professionally controlled model of learning and teaching across the institution that when studied or showed in the course of normal work, a distinctive value proposition for students. The value proposition would be clearly evident in the quality of teaching at scale and causally connected to the amount and quality of learning and/or the time taken to learn.

The longitudinal research on the effects of universities on student performance shows that such an effect is not the case. These studies show that most of the variance in student performance is predicted by their characteristics at entry. The gains in performance from entry to graduation are small and there is much more variability in student performance within universities than between them indicating the lack of variance contributed by a university's approach (ACT 2009; Liu 2008; Pascarella and Terenzini 2005). Further, while progress and graduation rate are employed as productivity outputs, we know that many students do not meet the expected rate of progress to graduation. The 6-year graduation rate at 4-year universities is 60% for degrees of 4-year duration (NCES 2016). We also know that when universities describe their efforts and intentions about learning and teaching the claims of effect are described in amorphous aspirational language that is astoundingly similar from institution to institution and lacks serious empirical support and validation (Dvorak and Busteed 2015; Hil 2015). According to Dvorak and Busteed (2015):

> The lack of enduring and unique identities in higher education offers an opportunity for education leaders, as it indicates there are a host of undifferentiated brands ripe for disruption. (p. 2)

Under these circumstances, it should come as no surprise that the contemporary determination of a great university seems to be predicated on the four R's rankings, research, real estate, and rah! (sports) (Craig 2015).

Not only do universities lack the models of practice at scale required to impact learning and teaching quality, they lack the internal analytic and evaluation systems, methods and tools to measure what they are doing in valid and reliable ways (Bain and Drengenberg 2016). This is not because of a lack of research and practical guidance. Extensive longitudinal research exists related to efficacious practice in learning design (e.g., Biggs and Tang 2007; Koper 2006), assessment (e.g., Sadler 2005; Wiggins 1998), and pedagogy (e.g., Hattie 2008, 2015; Marzano 1998), and approaches for implementing protocols derived from these practices. The issue is whether universities have the capacity to design themselves in ways that employ an understanding of the learning and teaching context to produce professionally controlled practice for the benefit of all faculty and students. To alter these circumstances requires an understanding of what learning and teaching means with sufficient clarity to establish differences or distinctions in the quality of practice at scale.

1.4 Interrogating the Effort Chain

What follows is an explanation of why the critical absence of professional control confounds each of the links in the effort chain described in Fig. 1.1 and renders each of limited effect in terms of a whole-of-organization impact on learning and teaching.

1.4.1 Standards

Most universities are governed by external accreditation standards designed to assure the scope and quality of what universities do. Standards from professional fields also govern what universities need to provide in order to meet the requirements of the professions while sector standards (state or national) address universities' capacity to deliver programs including things like admission expectations, faculty qualification, and in some instances the quality of teaching and assessment. These standards are intended to guide both the content and delivery of university programs.

Whether at the sector or professional level, these governing standards falter because they lack underpinning evidence-based protocols for visible and comparable practice that would give meaning to them. For example, the following standard for course design is from a national body responsible for the accreditation of universities "Assessment tasks for the course of study and its units provide opportunities for students to demonstrate achievement of the expected student learning outcomes for the course of study" (TEQSA 2015, p. 16). The utility of the standard is predicated upon a shared underpinning of quality assessment and a sense of what constitutes an appropriate outcome. By shared underpinning we mean that the standard reflects protocols or guidelines regarding what quality assessment practice looks like and what students should be expected to learn in relation to that assessment practice. However, we know that comparable and visible evidence-based professional knowledge and practice of quality assessment does not exist at scale in universities. Under current circumstances, any conceptualization of assessment could meet this standard as long as it could somehow show a defensible connection between what is taught and what is measured rendering the standard ineffective. There is no sense that this standard could drive a workable distinction in the quality of assessment in routine practice.

A second example is drawn from a program of national professional standards for middle school teachers: "Middle level teacher candidates demonstrate their ability to analyze content and assess the needs of their students. They purposely select and integrate teaching and assessment strategies that include information literacy skills and state-of-the-art technologies for all students." (AMLE 2012, p. 1). We see the same problem at play in this example as those described for the sector standards. There is no accepted body of practice to guide a determination about what constitutes *teaching and assessment strategies* and as a consequence there is no specific guidance for a university faculty of education regarding the scope and quality associated with the way they design a program to prepare their preservice school teachers to meet the standard. Without the detail provided by specific research-derived protocols for practice, both of the standards described in the examples are unable to make the valid and reliable distinctions necessary to discern quality.

The following statement is from the standards for approving a general practice in medicine: "Our practice has a consistent approach for the diagnosis and management of conditions affecting our patients in accordance with best available evidence" (RACGP 2012, p. 32). On face value, the standard could be viewed as even more general that the previous exemplars. What makes this standard different is not so much the level of detail in the narrative describing the standard but the way it is linked to best available evidence (i.e., clinical guidelines) in a range of sub-fields (e.g., pharmacy, oncology, etc.) that instantiate the standard. The generic description of the standard is underpinned by and has clear line of sight to professionally controlled evidence-based protocols for practice that are missing in standard setting for learning and teaching in higher education.

When challenged with the argument about the lack of professional control, standards bodies and many universities often respond with claims about the dangers of standards being overly prescriptive, the importance of flexibility, respecting autonomy, and/or the exceptionalist nature of learning and teaching work. We contend that these claims show a basic confusion about, and lack of understanding of, the most basic features that define a professional field and especially in relation to the foundational role of protocols for professional practice.

1.4.2 Quality Assurance Through Governance

Many universities possess internal policy and organizational frameworks for assuring the quality of the learning and teaching experiences they provide to students. These efforts focus on regulatory/audit requirements for the design of programs and courses, and committee-based reviews convened to ensure that the offerings meet external sector or professional standards. Policies frequently set in place the specific design features required for program and course approval and the process and procedures to implement the policy in practice. What we know about these policies and governance processes is that like standards, they are not informed specifically by evidence-based protocols for practice and are largely uninformed by theory and research derived from professional knowledge about learning and teaching (Henard 2009, Hora 2012; Cochran-Smith and Zeichner 2005; Hoban 2004; Kezar and Lester 2009). According to Henard (2009):

> The vast majority of initiatives supporting teaching quality are empirical and address the institutions' needs at a given point in time. Initiatives inspired by academic literature are rare. (p. 6)

Governance practices focus on mitigating organizational risk over curricular innovation or best practice by amplifying institutional requirements for efficiency and compliance that lack any known impact on student learning (Coate and Tooher 2010). For example, governance processes are frequently built around the kind of standards we just described that lack a foundation of visible and comparable evidence-based practice. Universities through their internal processes claim compliance and risk mitigation by invoking governance structures and processes referenced to the standards. However, these processes are incapable of driving genuine quality control and assurance and more importantly improving the quality of learning and teaching at scale because as we have seen, the standards to which they are referenced are incapable of doing so. Reviews of policy and organizational frameworks show a profound absence of empirically derived research as a term of reference in the governance frameworks for assuring the quality of learning and teaching. There are high levels of contestation about quality and limited reference to practices that improve learning in the policy and administrative requirements that guide program and course design (Jarvis 2014; Zundans-Fraser and Bain 2015).

Not surprisingly, there is no body of evidence showing a discernable effect of quality assurance approaches on learning and teaching in higher education (Ewell 2010; Harvey and Newton 2007; Stensaker 2008). Studies of the effects of quality assurance efforts show that it may influence the power structure, transparency and public relations of an institution yet has little to say about quality learning and teaching (Stensaker 2008). In effect, governance efforts create a quality illusion where universities are busy with the work of creating regulations, assuring compliance, and mitigating risk, although all they really do is amplify the fundamental lack of professional control in the standards of practice they are seeking to meet. Despite the busyness created for leaders in universities designing, building and implementing governance processes and frameworks, they have not been shown to make a difference in the quality of the learning experience or in the achievement of student outcomes.

1.4.3 Evaluation of Teaching

The lack of professional control echoes in the survey approach to the frequently controversial student evaluation of university teaching. Student evaluations remain the primary source of evaluative feedback about learning and teaching in universities. Students are routinely asked to evaluate faculty members on the quality of their practice including quality of instructor feedback, clarity and quality of teaching, quality and use of resources (Marsh 2007). The answers students provide to survey questions often have high stakes implications for faculty in terms of promotion and termination; and for the institution when they are aggregated and used to rank universities according to the quality of the learning experiences students can expect if they attend a particular institution. In each case, (i.e., faculty and university) student responses are framed as measures of quality when they are clearly at best measures of satisfaction especially when the results are aggregated as an indicator of a university's overall performance. Despite the high stakes use of student evaluations as measures of quality, there have been no whole-of-organization studies that evaluate teaching using professionally controlled practice as a criterion variable.

To be a valid measure of quality, student responses would need to reflect expectations about known comparable and visible practice occurring in classrooms and online course experiences. The survey items would reflect a university's model of teaching, the things it believes, values, and does at scale. Those beliefs, values, and actions would be established in practice in comparable and visible professionally controlled ways and communicated to students. Items included on the evaluation survey would reflect those values, beliefs, and practices (e.g., whether assessment tasks included explanations of criteria for success, or whether intended outcomes, content, and assessment were aligned specifically in the design of learning experiences). Students would be asked questions about the presence or absence of these features in their learning experiences. The measurement devices would possess both construct and content validity. Under professionally controlled circumstances, student feedback would serve as an important reliability check for a known model of practice including the use of well-supported pedagogy, feedback methodologies, and the match between pedagogy, resources, and content. This is almost never the case. Students are more frequently asked to respond to items about whether teachers showed concern for them, were enthusiastic, and motivated them to learn. This is analogous to asking patients about their doctors' bedside manners without attending to the quality of their medical practice. When answering such questions, students are expressing satisfaction with a loosely defined articulation of a largely idiosyncratically constructed learning experience. Under these circumstances, students respond with what they know given the autonomous and idiosyncratic ways teaching is/was constructed in their own experience as learners and without reference to those things that are known to produce better learning and teaching outcomes.

It is reasonable to expect that satisfaction expressed in surveys would in some way covary with quality, although this can only be established validly and reliably within a context of professional control. Satisfaction is no guarantee of quality under current circumstances. For example, consumers can be satisfied with a service/product that does not work as claimed or at all, is unnecessary, or performs poorly, if at a given point in time it represents state-of-the-art knowledge or is the subject of a successful marketing and sales effort. The history of all fields is replete with examples of this phenomenon (e.g., bloodletting in medicine, bottled water, snake oil in pharmacology). Student evaluations in universities are an expression of satisfaction with an experience that does not reflect evidence-based professional control at any scale beyond the practice of an individual teacher.

While the preponderance of the literature frames the many issues associated with student evaluation as symptomatic of more nuanced measurement concerns (Drengenberg and Bain 2016), we contend these issues are a symptom of the more fundamental validity problem described here. If an institution has not assigned value to what it means by pedagogy, feedback, appropriate resources, etc., (a model of learning and teaching), and then seeks to evaluate quality teaching, the best that can be expected from such an effort is a determination of student satisfaction with whatever a faculty member believes to be good teaching practice. Whether this lines up with anything we know about effective teaching is unexplored in evaluation regimes. Further, and of greater importance, without knowledge of professionally controlled practice at scale, it is simply not possible to generate the workable distinctions in practice necessary to infer the differential quality of a faculty member's work or the performance of the institution as a whole from the results of such an evaluation. For example, determining whether a given teaching approach, assessment practice or learning strategy is being employed more or less effectively is only possible when such approaches are defined according to known protocols and accepted standards of practice. These distinctions must be understood to discern quality and address issues and problems.

1.4.4 Promotion Frameworks

Many universities are sharpening their expectations and requirements for promotion as part of their effort to prioritize and professionalize teaching (Norton, Sonnemann, and Cherastidtham 2013) with the goal of creating workable distinctions among candidates for promotion. Most academic staff view promotion as the primary source of recognition for the quality of their contribution in the teaching domain (Probert 2015). Figure 1.2 describes an excerpt from a faculty learning and teaching promotion framework in the area of *program design and development*. It shows

Domain	Assistant Professor	Associate Professor	Professor
Program Design and Development	Peer Review of program & course design outlines and resources Documented assessment tasks indicating variety of approaches & authentic tasks. Adoption of teaching resources by others Learning & teaching grants Mapping and integrating quality assurance frameworks & standards	Adoption of teaching resources by other universities Peer reviewed grant reports National Fellowship	Adoption of teaching resources by international universities

Fig. 1.2 Excerpt from an academic promotion framework for learning and teaching

requirements at three levels of sophistication with the level of demand increasing from left to right on the figure in line with the different ranks of a candidate for promotion. The excerpt can be employed to illustrate the way a lack of comparable and visible professionally controlled practice also serves to undermine the intent of bringing more rigor and focus to the role of teaching in academic promotion. It is broadly representative of the way many promotion frameworks are constructed.

The example is illustrative of a number of concerns. First and foremost, none of the excerpts are capable of producing the kind of workable distinctions required to discern quality because like the standards, governance models, and evaluation of teaching discussed earlier, they are not underpinned by comparable and visible professionally controlled practice in the areas described. For example, some questions about the excerpt could be: What would determine a successful review of a course design outline from an unsuccessful effort? What kind of variety in assessment approach would be successful? What criteria would be employed to determine the adoption of resources by others? Genuine answers to these questions rely on evidence-based guidance implemented at scale. This is rarely if ever the case in normal learning and teaching practice. With all likelihood, any scholarly well-developed presentation when combined with the connoisseurship-type imprimatur of a promotion committee or panel would suffice to meet the standard described in Fig. 1.2 irrespective of the evidence base for, or efficacy of, the practice or approach presented. In this case, connoisseurship becomes a proxy for the lack of real standards by which to judge performance.

Figure 1.2 also provides evidence of another phenomenon common in the current practice of learning and teaching in higher education, the use of external peer validation as a substitute for workable distinctions in practice as evidence of quality. We see in the examples described in Fig. 1.2, that if someone else will adopt a process or practice developed by a candidate, or if a candidate is invited to make judgments about the work of others their standing and that of their work is then deemed to be of quality. The wider the adoption (i.e., international universities, national boards) or the more prestigious the invitation the more recognition and status accorded to the candidacy for promotion. Those who succeed in this model become the adjudicators for new applicants in an infinite loop of connoisseurship without professional control. This kind of approach is also employed at scale among universities to benchmark their learning and teaching practices (e.g., the university assessment model) and meet standards (e.g., Chalmers 2011; Dunn et al. 2007). We only have to reflect on the extant research, cited earlier, the absence of professionally controlled practice across higher education, including the lack of valid and reliable determinants of quality across the sector to see the limitations of this approach. We also know that the general level of knowledge of professionally controlled practice is low in many universities (e.g., Scott and Scott 2015) which calls into question the qualification of at least some of those doing the peer reviewing. In reality, consensus-driven connoisseurship absent professional control is a poor substitute for understanding the learning context in ways that are capable of producing genuine workable distinctions in performance and quality.

The content of Fig. 1.2 also shows how promotion is linked to other common elements of the evidence chain. In this instance receiving a fellowship from an external agency or center (i.e., National Fellow). The assumption here is that there is a relationship between the construct of learning and teaching competence as conceived for the purposes of promotion and receiving an award from a national body, again without linking the award to evidence of visible and comparable standards of professionally controlled practice. Being the recipient of an award from a recognized body may seem to be compelling evidence of excellence and impact. However, the validity of the award can only be made clear when it is aligned with the construct measured in the promotion framework and based upon verifiable evidence of performance in relation to professionally controlled practice. The issue here is not about the general value of an award or its significance but more what it actually means and further, the extent to which it reflects or predicts the intent defined in a promotion framework in a field where there is so much ambiguity about standards, efficacy, and quality.

1.4.5 Centers for Learning and Teaching

Much of the work related to the initiatives described in the effort chain fall under the responsibility of centers of learning and teaching. Whether, within institutions or elsewhere (e.g., state and national programs), centers or institutes for learning and teaching are charged with providing expert advice, building faculty capacity, assuring quality, and conducting awards and grant programs across the university. From their inception, they have come to represent the most common approach to improving learning and teaching in higher education (Cross 2001; Singer 2002). Singer (2002) describes these centers as the hubs of educational reform in universities. They create opportunities for faculty members, disseminate research, provide professional development, and support the process of responding to data derived from student evaluations of teaching. These centers frequently assume responsibility for the quality of online learning and learning analytics. Centers are also responsible for realizing the strategic intent and outcomes of the university's efforts to promote learning and teaching and build capacity. They are mandated to have a scalable impact. In their efforts to make a difference they inherit all of the problems that face the links in the effort chain standard setting, governance, promotion frameworks, and evaluation efforts.

The fundamental problem: Their work does not occur within a context of professionally controlled practice that would make possible the scalable influence that is frequently at the core of their mandate. Those who lead these centers are required to provide high-level advice and leadership although their roles rarely extend to creating the kind of context, the fundamental structural change required to realize the aspirations their organizations have for the centers they lead.

While there may be an intent to make universities better at learning and teaching as a whole, centers for learning and teaching are designed for something else: To influence learning and teaching through elective, selective, and exemplary approaches that are incompatible with whole-of-organizational change. By elective, selective, and exemplary, we mean approaches that offer elective capacity building and support for promotion; recognize excellence selectively through awards and grants; seek to demonstrate impact by example as opposed to scalable systemic effect, and where all of these actions occur within a context free of professional control. The expectation seems to be that exemplary practice sponsored by such centers will somehow exert a kind of osmotic effect on the whole organization which never happens at the scale required to influence quality overall.

Without models of scalable practice, centers become supporters and guides on the side for good works and better practice, identifying and recognizing academics who pursue excellence, providing and facilitating grants for those who are trying to innovate, participating in the development of policies, frameworks, producing papers and disseminating resources about effective practice and assuming roles and responsibilities in the existing governance processes. Even when initiatives have a systemic mandate or intent (e.g., institutional policy development; whole of faculty capacity building) they are stymied by the lack of a university-level guiding understanding of context required to focus and situate the efforts in ways that make genuine quality control possible at scale. The lack of whole-of organization impact associated with the effort chain initiatives and the role of centers in leading that work often results in diminished uptake of their offerings by faculty members (e.g., higher education teaching certification, Probert 2015) who are frequently skeptical about their value.

At state and national levels, the pattern of engagement is the same, higher level awards, national grant programs, and capacity building initiatives, all possibly more competitive and prestigious than their local analogs but lacking the same foundation of valid professionally controlled practice required to fulfill their mandates and the intents of the policy makers who fund them. These may be strong and somewhat disconcerting claims, however if we revisit the research cited earlier about the inability of universities to articulate their learning and teaching practice, to distinguish themselves from peer institutions, exert a clear demonstrable effect on student learning at scale, along with the unremarkable history of quality assurance, they are far from being overstated or unreasonable.

1.4.6 Elusive Productivity

It may seem ironical and/or possibly no surprise given the picture painted here that learning and teaching productivity has become an important focus of the higher education conversation. Questions about the value and performance of higher education are increasingly driving efforts to establish just how productive universities and colleges are in their engagement with learning and teaching (e.g., Chang et al. 2014). Best practice measures of learning and teaching productivity (e.g., Sullivan et al. 2012) have attempted to apply approaches used in other fields to provide higher education with a standard approach to measuring the productivity of their learning and teaching efforts. The best example is the Tornquist Index (discussed in detail in Chap. 8) that examines change over time in an index of inputs and outputs associated with learning and teaching. Inputs of time, use of resources, and capital infrastructure and outputs of program and course graduation are employed in the index (Sullivan et al. 2012). The approach assumes a relationship between the expenditure of time, consumption of resources and course and program completions despite acknowledgement by the authors of the inability of the field to determine quality. As described earlier, they acknowledge that methods like price competition and performance studies employed by other fields do not translate well to determining the quality of learning and teaching in higher education (Massy et al. 2012). Discerning quality is seen as a measurement problem from a productivity perspective although as has been illustrated throughout this chapter the problem is much deeper and more pervasive. The quality problem is a symptom of the inability to understand what learning and teaching means in professionally controlled ways. It is clear that given the ambiguity associated with learning and teaching, there should be profound concerns about measuring productivity using input measures of time and resource consumption and program and course graduation/completion as outputs.

The contribution of universities to learning, the quality and productivity of their work and their understanding of quality is at best murky with many more questions than answers about the relationship between what universities do and their impact on the learning experience of their students. At this point in time, no university has the capacity to determine the effect of its learning and teaching aspiration on student outcomes at scale. In higher education, the conduct of learning and teaching are at best an approximation and more an impersonation of the kind of professionalism we see in other fields like medicine, engineering, and law where longstanding models of practice apply, and where as a result quality can be discerned on the basis of differential effect or outcome.

The existence of accreditation standards, awards for excellence, evaluation scales, criteria for promotion all imply underpinning models, and bodies of scalable practice that would legitimize these initiatives. However, they do not exist, rendering such efforts and initiatives illusory as indicated by research on both quality assurance efforts and the student outcomes of higher education. The effort chain is employed quite deliberately by universities to impersonate fields with genuine professional control. In reality, the chain drives an infinite loop of self-reinforcing and self-serving crude approximations of learning and teaching quality (e.g., institutional benchmarking) incapable of making valid workable distinctions in practice. Many universities have offices specifically dedicated to the assurance of quality. They are charged with gathering the available data on teaching and learning, monitoring attrition and retention, benchmarking, conducting focus groups, and reporting internally and externally. These offices also amplify the problems identified in the effort chain because they promote the limited understanding and conceptualization of learning and teaching by organizing and disseminating weak data derived from the loosely coupled approaches described here.

1.5 Understanding Context

In the preceding sections, we have shown why an understanding of context is currently not possible given the absence of professional control and loose coupling (Weick 1976) of initiatives currently employed in colleges and universities to promote and assure the quality of learning and teaching. We contend that to address the problems associated with the current effort chain approach, we need to reconceptualize the challenge by focusing on the context within which learning and teaching occurs. Figure 1.3 describes a cycle of conditions to create a genuine professional context for learning and teaching in higher education and ultimately determine the quality of that work. The cycle expresses the interrelationships that are necessary for a university to develop a successful, scalable approach to learning and teaching.

1.5.1 Models of Learning and Professional Practice

An understanding of context begins with a model of learning and teaching. It is in this stage that a university seeks to link its thinking, beliefs, and values to existing theory and research-based conceptual models. The majority of professional fields have progressed by proposing and testing models and specific practices, discarding those that fail to adequately account for successful outcomes and instantiating those

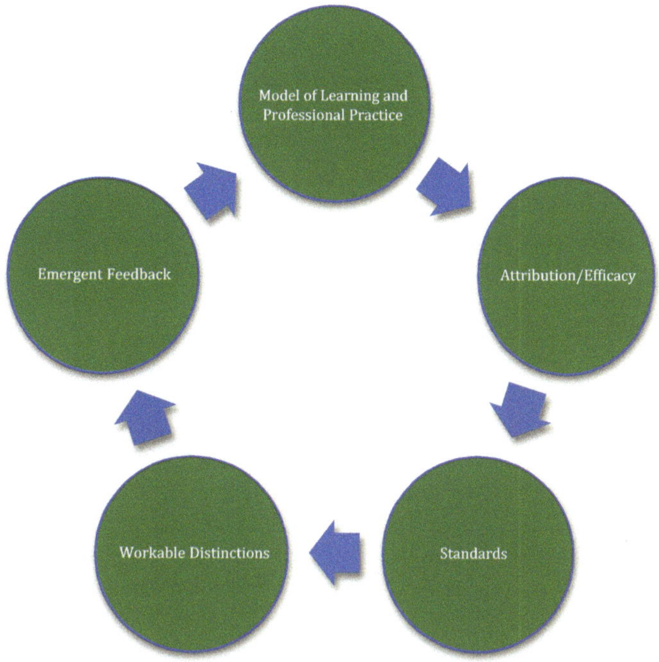

Fig. 1.3 The learning and teaching context cycle

that do in a dynamic and sometimes haphazard developmental trajectory (Kuhn 1996; Bain and Drengenberg 2016). In education, models are the entry point for defining context. They explain the way aptitude, instruction, affective, and environmental factors influence learning. This includes longitudinal research undertaken by Bloom (1976), Creemers (1994), Entwistle (2000) and Walberg (1986). While much of this work has been done in the compulsory sector, each of these researchers has produced an empirically derived understanding of the relationship between key factors that influence learning and the way they interact. They can enable a university trying to understand its learning and teaching context to map at a high level the territory of learning and teaching.

A determination of professionally controlled practice instantiates the model and begins to articulate what the conceptual framework looks like in normal practice. This element of the context cycle defines the practices and approaches the university will use to translate its conceptual approach to the methods and protocols used in normal work. For example, Bloom (1976) found that approximately 25% of the variance in student achievement was explained by the quality of instruction. However, the model does not fully explicate what instructional quality means. The professionally practice component or step of the cycle involves identifying those evidence-based comparable and visible practices that will express in practice what quality of instruction actually means. To extend the example, cooperative learning

(CL) is a teaching approach that has an extensive research history as an effective practice at all levels of education including higher education (Springer et al. 1999). CL is one example of a practice that could be used to instantiate a model of learning and teaching. When a university employs CL (among other approaches) it can contribute to the quality of instruction and produce significant learning effects when the practice is implemented with integrity (Hattie 2015, 2008). The step of instantiating the model involves selecting those evidence-based approaches that turn a conceptual framework into a design for normal work and includes practices for pedagogy, student engagement, curriculum design, and assessment all applied at scale.

1.5.2 Attribution and Efficacy

A key feature of any learning and teaching context is evidence of an attributable relationship between the model of teaching and learning, and student outcomes. This is a relationship that we can expect to be efficacious because the practices and approaches employed to instantiate the model have a preexisting track record of success established in empirical research. By instantiating its model with professionally controlled evidence-based practices, a university takes two major steps forward in understanding its context. First, it engages with approaches that can be expected to result in an attributable and efficacious relationship between what teachers do and student learning when deployed at scale. Second, it brings definition to what teaching and learning mean creating the conditions for the performance feedback and study required to discern quality at scale within an institution. Practices like CL among many others possess extensive guidance about how to design and deliver learning experiences that include the key features of the practice (e.g., structuring interdependence and individual accountability) (Slavin 1996), making visible and comparable what a given teaching practice means. The educational productivity research includes an extensive compendium of practices that include guidance (protocols) for implementation and evidence of their efficacy (Hattie 2008, 2012; Marzano 1998).

1.5.3 Professional Standards

When a model of learning and professional practice includes protocols for comparable and visible evidence-based practice, professional standards become discriminative because they are imbued with meaning and as a consequence can make a meaningful contribution to understanding the learning context. They can produce workable distinctions in practice. We can return to the assessment example described earlier to illustrate this point. We indicated that the standard —"Assessment tasks for the course of study and its units provide opportunities for students to demonstrate achievement of the expected student learning outcomes for the course of study" (TEQSA 2015, p. 16) was rendered ineffective because of a lack of underpinning professional control related to assessment practice. However, if a comparable and visible professionally controlled approach is taken to assessment, a university can make an assessment standard capable of driving quality practice. For example, a university may choose to employ a criterion-based approach that includes clear guidance for describing and contextualizing assessment tasks by linking it to instruction and clearly defining criteria for different levels of performance (e.g., Wiggins 1998). Such an approach can turn a standard from generic aspiration to a rigorous evaluable protocol because the core construct of assessment has an underpinning meaning expressed in professionally controlled practice required at scale across an institution.

1.5.4 Workable Distinctions

The interaction of model, professionally controlled practice, and standards create the conditions whereby workable distinctions in the day-to-day practice of learning and teaching become a realistic expectation (Drengenberg and Bain 2016). The things that matter for student learning can become the focus of feedback that determines whether an approach is enacted more or less effectively. We can see whether assessment, cooperative learning, or any other practice is being done well. When students are asked to provide feedback on assessment or pedagogy a university that understands its context knows what those things mean, and the key features of effective practice that possess an attributable and efficacious relationship with student learning. Further, a student who experiences professionally controlled practice on a consistent basis will come to know about the key success characteristics of those approaches. Their feedback will be informed by that knowledge and understanding. Under these circumstances, it becomes possible to use student evaluations as *a* although not *the* singular source of evidence for making distinctions between more or less effective practice that are both valid and reliable.

1.5.5 Quality at Scale

When a university's learning and teaching model is instantiated with professionally controlled practice, produces an attributable and efficacious relationship between learning and teaching based on workable distinctions in normal practice, it becomes possible to discern quality. For example, a determination of whether the known elements of quality cooperative learning or criterion-based assessment are being implemented in face-to-face or online learning can be made at all levels of the organization in ways that are valid and reliable. This happens because the context cycle described in Fig. 1.3 is understood and produces a clear line of sight from big ideas about learning to comparable and visible day-to-day practice. The community is able to reflect on its practice in professional ways focusing on features that are known to influence learning. Students can be asked about professionally controlled learning and teaching.

A university that understands its learning and teaching context in this way possesses the capability to complete the kind of work that is analogous to the performance studies and competition that underpins productivity models in other fields. If for example, the university has a known model instantiated in practice it can work out how long the process of designing programs and courses should take and what should be included. The standards of practice required in the normal work of design and implementation are made clear. It becomes possible to know whether programs and courses are designed to comparable and visible professionally controlled standards and to distinguish more or less effective practice, a point at which quality can be determined. When the context is understood, the use of time as a measure of productivity becomes viable because time is underpinned by a rich understanding of what learning and teaching mean.

1.5.6 Emergent Feedback

When the learning and teaching context is understood, emergent feedback (Bain 2007) can become part of that context. As described earlier, students (and faculty) who understand professionally controlled practice can provide feedback about the practices in the model. Instead of providing general feedback on satisfaction, they can be asked whether assessment rubrics were clear, whether there was sufficient structure in the cooperative learning tasks, etc. Feedback emerges all of the time from multiple sources at each point in the cycle of design, enactment and engagement with learning and teaching. Because the model is visible and comparable, we do not have to wait until after the fact to get feedback, it emerges at all stages in the design, delivery of, and engagement with, programs and courses. After the fact, judgement is replaced by the agency of all who are involved and have a stake in the learning and teaching work of the university. This kind of all-of-the-time feedback results in a dynamic system where the model, the practices, standards, and even the feedback mechanisms exist in a state of ongoing change.

The key distinction between the loose coupling of the effort chain and the context cycle is a systemic and systematic approach to learning and teaching. Practices like criterion-based assessment, cooperative learning and a host of others are part of the current conversation and process in many universities. Unfortunately, those practices are unlikely to realize their potential benefit for faculty and students unless they become part of a professionally controlled context underpinned by a model of practice and protocols that are capable of scalable implementation and impact. Universities that adhere to an effort chain approach can be described as *pre-contextual* (Bain and Drengenberg 2016). They are yet to understand and develop a professional learning and teaching context at scale. The context cycle described earlier in Fig. 1.3 can be viewed as a destination for a university seeking to exert professional control over learning and teaching. It represents, at a conceptual level, what needs to be accomplished in learning and teaching to exert a whole-of-organization effect.

1.6 The SOU: Transforming the Pre-contextual University

The SOU is an approach for creating professionally controlled scalable change that migrates universities from pre-contextual state to one where they can influence the quality, efficiency, and productivity of learning and teaching at scale. It is predicated on an assumption that learning and teaching process must be embedded at scale in the design of the organization. The design pivots on an understanding of context and professional control that can exert systemic effect. In the chapters that follow, we will show how to apply six design elements to make the context cycle a reality in the day-to-day practice of a university. This includes how to mobilize the community in order to build a model of learning and teaching; to identify practices and embed them in the model; to organize the governance of learning and teaching for quality outcomes; agency and the role of leaders; to build the technologies required to empower faculty and students with new sources of agency, and generate the feedback required to create and sustain a SOU. Figure 1.4 describes the elements of the model. Its form emphasizes the way in which each element is embedded in all others.

Following, is an introduction to the SOU, its design elements and underpinning principles. Each design element, introduced in Fig. 1.4 is the subject of a chapter in the book.

1.6.1 Committing to Learning and Teaching

In Chap. 2, we employ a central principle from complex systems theory to show how universities can establish the core beliefs and values required to build a model of practice for learning and teaching. We use contrasting examples to distinguish this approach from mission statement development and strategic planning efforts showing why these approaches are unlikely to produce transformational change. We show how developing simple rules or commitments drive model development and the ways to mobilize a community to develop and disseminate both the commitments and a model.

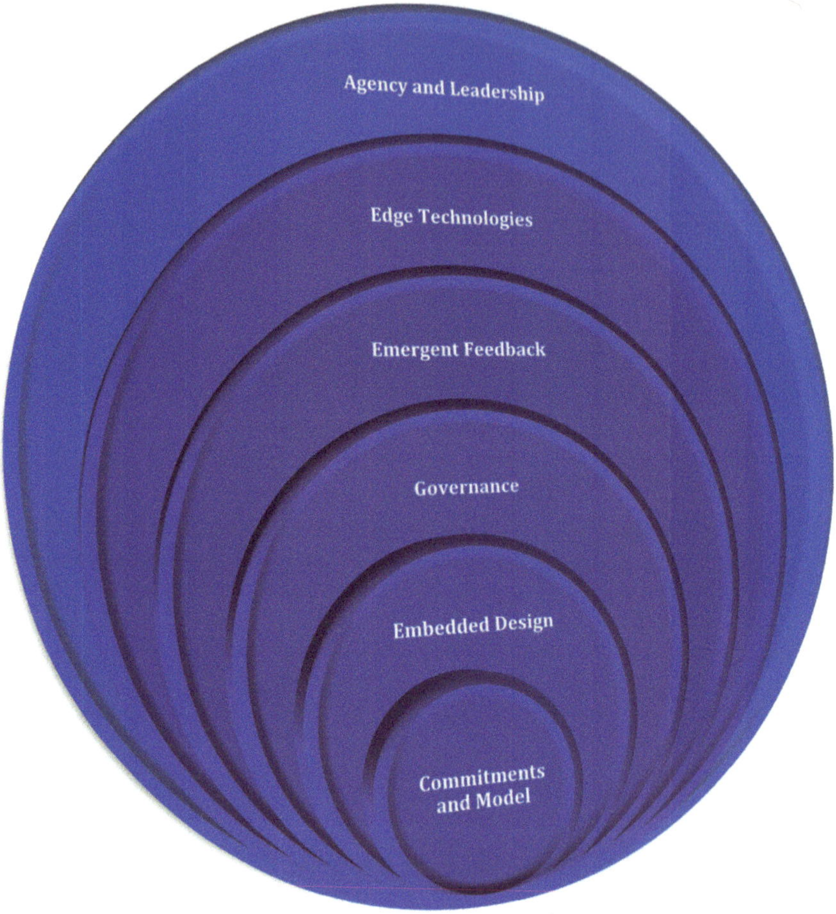

Fig. 1.4 Elements of the Self-organizing university

1.6.2 Embedded Design

Chapter 3 shows how to translate a conceptual model into practice by instantiating the elements of the model with evidence-based approaches to learning and teaching. We provide examples of evidence-based practice and the way they are represented in the design of the organization. We use the commitments and model to connect the practices together in ways that create a guide for the normal work of faculty and leadership. We then show how to confirm commitments, model, and practices in the design of an SOU. Examples included in this chapter compare the current effort chain approach with a context cycle method to embed professionally controlled practice that defines the normal whole-of-organization work of learning and teaching.

1.6.3 Emergent Feedback

In Chap. 4, we show how the articulation of model, practice, policy and organizational design creates the conditions for powerful new forms of emergent feedback. In the SOU, feedback emerges all of the time from the context cycle and the way learning and teaching is made visible and comparable in day-to-day work. We show how feedback happens while courseware is being designed when it is delivered and in terms of student performance. This emergent ground level feedback can then be analyzed at any level of the organization, program, school, faculty, and university. The chapter employs examples of feedback to show the workable distinctions required to discern quality in learning and teaching. The feedback made possible by the SOU provides a university with the evidence base to substantiate its unique value proposition.

1.6.4 Governance

In Chap. 5, we compare the governance approach in an SOU to the process in a pre-contextual university. We explain how a network of self-similar teams at different levels of the management structure of an SOU creates a flatter less hierarchical governance structure. We describe how the agency of stakeholder groups is represented in this system and use comparative examples to highlight the differences between a hierarchical and network approach to the university's organizational design. The chapter includes an example of the network of teams at faculty and university level and the way the self-similar design of the teams and their process enables them to interact.

1.6.5 Agency and Leadership

In Chap. 6, we describe what agency and leadership looks likes in the SOU. We explain how and why agency in higher education has evolved without an understanding of professional practice and how the learning and teaching power of a university can be transformed when faculty at all levels of the university possess a shared schema for learning and teaching. The chapter shows what the promotion process looks like in a self-organizing university and compares the approach with the effort chain model of promotion frameworks described in Chap. 1.

1.6.6 Technology for Learning and Teaching

In Chap. 7, we describe the learning and teaching technology system for a SOU including examples of what those tools look like in practice. We explain how an understanding of context creates the conditions for a new approach entitled Edge technology (Bain and Weston 2011) that reflect the context and leverages emergent feedback. Edge technologies extend the agency of users, enable collaboration and improve the network capability of the organization by shortening the edges or distances between those in the learning and teaching network.

This means if an organization adopts a particular visible and comparable approach to design its curriculum or specific pedagogical strategies (e.g., cooperative learning or criterion-based assessment), the key features of those approaches and especially those that influence achievement can be designed into the technologies used across the community. So, the tools people use in normal work can help them design and deliver courseware in ways that make it easier to incorporate comparable and visible evidence-based practice. The design of the tools helps users to employ professionally controlled practice. In this way, the tools have a virtual form of agency independent of their users (Levy 2001). Their design and functionality are a mirror of the university's learning and teaching model and context. They focus on those things that produce workable distinctions and quality. When employed within the SOU approach, they also revolutionize the concept of learning analytics (Bain and Drengenberg 2016). In Chap. 7, we explain the problems of the effort chain approach for learning analytics and how edge technologies create the opportunity to build a learning analytics system focused on the learning and teaching context and the feedback derived from professionally controlled practice.

1.6.7 Productivity

In Chap. 8, we present a powerful new model of productivity showing what productivity looks like when a university understands its learning and teaching context. We show how this understanding can transform the fiscal competitiveness of universities and the implications for the cost, efficacy, and value of a university education.

1.6.8 Moving Forward

What are the shifts in thought and action required to create self-organizing universities? Chapter 9 answers this question identifying the shifts required to move forward with the principles and practices of a SOU. We describe the forces for change and equilibrium and where thinking needs to go to make a SOU possible.

In summary, the purpose of this chapter has been to point to an *inconvenient truth*[1] about the way learning and teaching occurs in universities and specifically the way a lack of professionally controlled professional practice renders the work undertaken in the name of learning and teaching illusory in terms of whole-of-organization impact. The effort chain described in this chapter is not capable of the whole-of-organizational change promised by standard setting, governance, evaluation, promotion frameworks, and professional support. We described those universities incapable of whole-of-organizational change as pre-contextual. To address the effort chain problem, we posited the need for an understanding of context and introduced the learning and teaching context cycle to explain what context means and show what is necessary to address the quality illusion. To conclude the chapter, we introduced the key elements of the SOU as a design approach focused on addressing the problem and designing a university capable of understanding its context through professionally controlled learning and teaching practice.

In the chapters that follow, we will show an alternative course of action. Each of the chapters compare and contrast examples of the pre-contextual and SOU highlighting what is required to take up a whole-of-organization approach to learning and teaching.

Takeaways

The key takeaways from this chapter are:

- A university that does not understand its learning and teaching context is pre-contextual. It cannot influence the quality of learning and teaching at scale;
- Contemporary research on the effects of a university education and quality assurance shows that universities are unable to discern quality and differentiate themselves in relation to their core activity of learning and teaching;
- The effort chain of good works and intentions to improve the quality of learning and teaching are stymied by a lack of professional control in learning and teaching;
- The term professional control denotes a process of dynamic management and evaluation of comparable and visible practice, as opposed to something that restricts or constrains;
- A lack of professional control confounds efforts to influence learning and teaching at a whole-of-organization level;
- Quality reflects a deep understanding of the learning and teaching context;
- The effort chain when combined with connoisseurship as a substitute for professional control creates a recursive loop of self-reinforcing action that does not improve learning and teaching at scale;

---

[1]A term popularized by the 2006 motion picture of the same title about Al Gore's environmental campaign. David et al. (2006)

- The SOU is a design method for understanding context and building a whole of university approach to learning and teaching;
- Expectations for whole-of-organization improvement in learning and teaching require a whole-of-organization approach to change;
- The SOU involves the application of a series of integrated design principles to building learning and teaching commitments, organizational design, feedback, agency and leadership, and the role of technology in a university.

References

ACT. (2009). *ACT report: Voluntary system of accountability learning gains methodology.* Retrieved from https://cp-files.s3.amazonaws.com/22/ACTReport_LearningGainsMethodology.pdf.

AMLE. (2012). *Association of middle level education standards.* Retrieved from http://www.amle.org/AboutAMLE/ProfessionalPreparation/AMLEStandards.aspx.

Bain, A., & Drengenberg, N. (2016). *Transforming the measurement of learning and teaching in higher education.* New York: Routledge.

Bain, A. (2007). *The self-organizing school. Next generation comprehensive school reforms.* Lanham, MD: Rowman & Littlefield.

Bain, A., & Weston, M. (2011). *The learning edge: What technology can do to educate all children.* New York: Teachers College Press.

Biggs, J., & Tang, C. (2007). *Teaching for quality learning at university.* Maidenhead, UK: McGraw-Hill and Open University Press.

Bowker, G., & Star, S. (2000). *Classification and its consequences.* Cambridge, Massachusetts: The MIT Press.

Bloom, B. S. (1976). *Human characteristics and school learning.* New York: McGraw-Hill.

Chalmers, D. (2011). *CADAD benchmarking guidelines.* Retrieved from http://www.hedg.ac.uk/meetings/documents/DeniseChalmersBenchmarkingGuidelines.pdf.

Chang, C., Farr, M., & Whigham, N. (2014, October). *How much should you pay for a university degree?* Retrieved from http://www.news.com.au/finance/business/how-much-should-you-pay-for-a-university-degree/story-fnda1bsz-1227104966674.

Coate, K., & Tooher, M. (2010). The Galway symposium on design for learning: Curriculum and assessment in higher education. *Teaching in Higher Education, 15*(3), 347–354. doi:10.1080/13562511003740924.

Cochran-Smith, M., & Zeichner, K. M. (2005). *Studying teacher education: The report of the AERA panel on research and teacher education.* Washington, DC: American Educational Research Association.

Craig, R. (2015). The problem with college rankings. Retrieved from https://www.higheredjobs.com/blog/postDisplay.cfm?post=704.

Creemers, B. P. M. (1994). Effective instruction: An empirical basis for a theory of educational effectiveness. In D. Reynolds, B. P. M. Creemers, P. S. Nesselrodt, E. C. Schaffer, S. Stringfield, & C. Teddlie (Eds.), *Advances in school effectiveness research and practice* (pp. 189–205). Oxford: Pergamon.

Cross, K. (2001). Leading-edge efforts to improve teaching and learning. *Change, 33*(4), 31–37.

David, L., Bender, L., Burns, S., (Producers), & Guggenheim, D. (Director). (2006). *An inconvenient truth [Motion Picture].* United States: Participant Media.

Drengenberg, N., & Bain, A. (2016). If all you have is a hammer, everything begins to look like a nail—How wicked is the problem of measuring productivity in higher education? *Higher Education Research & Development.* doi:10.1080/07294360.2016.1208640.

Dunn, D. S., McCarthy, M. A., Baker, S., Halonen, J. S., & Hill, G. W. (2007). Quality benchmarks in undergraduate psychology programs. *American Psychologist, 62*(7), 650–670. doi:10.1037/0003-066X.62.7.650.

Dvorak, N., & Busteed, B. (2015, August, 11). It's hard to differentiate one higher-ed brand from another. *Gallup Business Journal,* Retrieved from http://www.gallup.com/businessjournal/184538/hard-differentiate-one-higher-brand.aspx.

Entwistle, N. (2000). *Promoting deep learning through teaching and assessment: Conceptual frameworks and educational contexts.* Retrieved from http://www.tlrp.org/pub/acadpub/Entwistle2000.pdf.

Ewell, P. (2010). Twenty years of quality assurance in higher education: What's happened and what's different? *Quality in Higher Education, 16*(2), 173–175.

Hattie, J. (2015). The applicability of visible learning to higher education. *Scholarship in Teaching and Learning in Psychology, 1*(1), 79–81.

Hattie, J. (2012). *Visible learning for teachers: Maximizing impact on learning.* New York: Routledge.

Hattie, J. (2008). *Visible learning: A synthesis of over 800 meta-analyses relating to achievement.* New York: Routledge.

Harvey, L., & Newton, J. (2007). Transforming quality evaluation: moving on. In D. F. Westerheidjen, B. Stensaker, & M. Rosa (Eds.), *Quality assurance in higher education: Trends in regulation, translation and transformation* (pp. 225–246). Dordrecht, The Netherlands: Springer.

Henard, F. (2009). Learning our lesson: Review of quality teaching in higher education draft report. Retrieved from http://www.oecd.org/edu/imhe/43961761.pdf.

Hil, R. (2015). *Selling students short.* Crows Nest, NSW: Allen & Unwin.

Hoban, G. F. (2004). Seeking quality in teacher education design: A four-dimensional approach. *Australian Journal of Education, 48*(2), 117–133. doi:10.1177/000494410404800203.

Hora, M. T. (2012). Organizational factors and instructional decision-making: A cognitive perspective. *The Review of Higher Education, 35*(2), 207–235. doi:10.1353/rhe.2012.0001.

Jarvis, D. (2014). Regulating higher education: Quality assurance and neo-liberal managerialism in higher education: A critical introduction. *Policy and Society, 33,* 155–166.

Kezar, A., & Lester, J. (2009). *Organizing higher education for collaboration: A guide for campus leaders.* San Francisco, CA: John Wiley & Sons.

Koper, R. (2006). Current research in learning design. *Educational Technology & Society, 9*(1), 13–22.

Kuhn, T. S. (1996). *The structure of scientific revolutions.* Chicago: IL, University of Chicago Press.

Levy, P. (2001). *Cyberculture.* Minneapolis: University of Minnesota Press.

Liu, O. (2008). *Measuring learning outcomes in higher education using the measure of academic proficiency and progress (MAPP).* Retrieved from https://cp-files.s3.amazonaws.com/23/RR-0847MeasuringLearningOutcomesUsingMAPP.pdf .

Marsh, H. W. (2007). Students' evaluations of university teaching: A multidimensional perspective. In R. P. Perry & J. C. Smart (Eds.), *The scholarship of teaching and learning in higher education: An evidence-based perspective* (pp. 319–384). New York: Springer.

Marzano, R. (1998). *A theory-based meta-analysis of research on instruction.* Aurora, CO: Mid-continent Research for Education and Learning. Retrieved from http://www.peecworks.org/peec/peec_research/I01795EFA.2/Marzano%20Instruction%20Meta_An.pdf.

Massy, W., Sullivan, T., & Mackie, C. (2012). Data needed for improving productivity measurement in higher education. *Research and Practice in Assessment, 7,* 5–15.

McCloskey, G. C., & Bulechek, G. M. (1994). Standardizing the language for nursing treatments: An overview of the issues. *Nursing Outlook, 42*(2), 56–63.

NCES. (2016). National Center for Educational Statistics. *Fast facts, Graduation rates.* Retrieved from https://nces.ed.gov/fastfacts/display.asp?id=40.

Norton, A. & Cherastidtham, I. (2015). *The cash nexus: how teaching funds research in Australian universities*. Grattan Institute. Retrieved from http://grattan.edu.au/wp-content/uploads/2015/10/831-Cash-nexus-report.pdf.
Norton, A., Sonnemann, J., & Cherastidtham, I. (2013). *Taking university teaching seriously*. Grattan Institute. Retrieved from http://grattan.edu.au/report/taking-university-teaching-seriously/.
Pascarella, E., & Terenzini, P. (2005). *How college affects students (Vol. 2): A third decade of research.* San Francisco: Jossey-Bass.
Probert, B. (2015). *The quality of Australia's higher education system: How it might be defined, improved and assured.* Office for Learning and Teaching Discussion Paper. Australian Government Office for Learning and Teaching. Available at http://www.olt.gov.au/resource-quality-australias-higher-education-system.
RACGP. (2012). *Standards for general practice* (4th ed.). Retrieved from http://www.racgp.org.au/your-practice/standards/standards4thedition/.
Sadler, R. (2005). Interpretations of criteria-based assessment and grading in higher education. *Assessment & Evaluation in Higher Education, 30*(2), 175–194. Retrieved from http://www.tandfonline.com/doi/pdf/10.1080/0260293042000264262.
Scott, D., & Scott, S. (2015). Leadership for quality university teaching: How bottom-up academic insights can inform top-down leadership. *Educational Management Administration & Leadership, 44*(3), 511–531. doi:10.1177/1741143214549970.
Singer, S. (2002). Learning and teaching centers: Hubs of educational reform. *New Directions for Higher Education, 119,* 59–64.
Slavin, R. E. (1996). Research on cooperative learning and achievement: What we know, what we need to know. *Contemporary Educational Psychology, 21,* 43–69.
Springer, L., Stanne, M. E., & Donovan, S. S. (1999). Effects of small-group learning on undergraduates in science, mathematics, engineering and technology: A meta-analysis. *Review of Educational Research, 69*(1), 21–51. doi:10.3102/00346543069001021.
Stensaker, B. (2008). Outcomes of quality assurance: A discussion of knowledge, methodology and validity. *Quality in Higher Education, 14*(1), 1470–1081.
Sullivan, T., Mackie, C., Massy, W., & Sinha, E. (2012). *Improving measurement of productivity in higher education. Panel on measuring higher education productivity: Conceptual framework and data needs.* Committee on National Statistics and Board on Testing and Assessment, Division of Behavioral and Social Sciences and Education. Washington, DC: The National Academies Press.
TEQSA (2015). *Higher education standards framework.* Retrieved from http://www.teqsa.gov.au/regulatory-approach/higher-education-standards-framework.
Walberg, H. J. (1986). Synthesis of research on teaching. In M. C. Wittrock (Ed.), *Handbook of research on teaching* (3rd ed., pp. 214–229). New York: Macmillan.
Weick, K. E. (1976). Educational organizations as loosely coupled systems. *Administrative Science Quarterly, 21,* 1–21.
Wiggins, G. (1998). *Educative assessment.* San Francisco: Jossey-Bass.
Zundans-Fraser, L., & Bain, A. (2015). How do institutional practices for course design and review address areas of need in higher education? *Higher Education Research and Development, 35*(4), 841–853. doi:10.1080/07294360.2015.1137883.

Chapter 2
Committing to Learning and Teaching

2.1 Introduction and Overview

In this chapter we show in detail the ramifications of the effort chain problem and explain why pre-contextual universities that employ this approach to learning and teaching are unable to produce transformational change. We describe how a lack of understanding of context and professional control (Bowker and Star 2000) severely inhibits the capacity of those pre-contextual universities to plan for and produce whole-of-organization improvement in learning and teaching. We attribute this problem to a mismatch between aspiration, organizational design, and agency. The chapter employs simple rules (Waldrop 1992) or commitments as a central principle of the self-organizing university (SOU) derived from complex systems theory, to show how universities can begin a change process in a different way by developing the core commitments that serve as the foundation for a systemic approach to learning and teaching. Crowdsourcing (Howe 2006) and Commons-Based Peer Production (CBPP) (Benkler 2002) are described as self-organizing methods for understanding the learning and teaching context, developing commitments and policy in the first phase of designing an SOU.

2.2 The Effort Chain in Action

In a book focused on learning analytics Bain and Drengenberg (2016) describe Conway's Law, which asserts the interface structure of a software system will show congruence with the social structure of the organization that produced it (Conway 1968). More recent empirical analysis of the law showed a clear congruence between the organizational structure of companies and their products (MacCormack et al. 2012). We contend the power of this law extends well beyond software development to the way an organization's existing structure defines and bounds the

A. Bain and L. Zundans-Fraser, *The Self-organizing University*,
DOI 10.1007/978-981-10-4917-0_2

scope and form of its strategic aspirations and capacity for change. For example, organizations most commonly use strategic planning processes to embark on major change. Strategic plans usually involve a mission statement, the formulation of goals and objectives in relation to mission, and a plan of implementation that identifies key actions, monitors progress, and evaluates outcomes (Hanover Research 2013; Martin 2014). Good plans are: flexible, well documented and well communicated, and explicitly represented in the management system (CFAR 2005). While good strategic planning encourages out-of-the-box thinking and the possibility of transformation as an outcome (Bertolini et al. 2015), we contend that the factor most likely to influence a strategic plan and its outcomes is the way an organization is constituted as it enters the planning process.

An organization's design is an expression of its schema—its beliefs and process. Beliefs and process are expressed in methods and tools, roles and responsibilities, how individuals contribute, and the way the organization as a whole addresses problems and develops more or less successful courses of action. Both the schema and those that work with it possess agency that functions in a reinforcing feedback loop. The organization's schema drives roles and responsibilities that shape individual perspectives on needs, priorities, and performance indicators. How this happens will affect schema development (Auhl, in progress). The schema and existing organizational structure are in turn reinforced by the efforts of those individuals. For example, it is common for the individuals most prominent in a change planning process to be the same people responsible for and most invested in the current condition of the organization. The organizational schema frames strategic planning aspirations, the capacity to act on those aspirations, and measure outcomes. If an organization's schema and structure and the capacities and dispositions of the individuals who work with that schema are incompatible with transformational change, the strategic planning process and those that work with the process will struggle irrespective of a stated intention to change an organization in transformational ways. Example 2.1 shows how this phenomenon of self-reinforcing action plays out in the reality of strategic planning for learning and teaching at Pre-contextual Grandbridge University.

Example 2.1 Pre-contextual Grandbridge University Develops its Learning and Teaching Strategic Plan

Pre-contextual Grandbridge University has been a national leader in distance learning (DL) for nearly three decades. It entered the DL field when few of its competitors considered DL to be of particular interest. The university filled a need by becoming one of a relatively small number of universities offering its core curriculum in a distance format. For over 25 years, a strong and consistent DL program secured the fiscal health of Pre-contextual Grandbridge and its distinctiveness as a leading DL provider. Over the last five years, circumstances have changed markedly. Now in the online and blended learning "business," Pre-contextual Grandbridge faces fierce competition across the sector from universities that 15 years ago viewed DL as a "watered down" form of higher education. Those same institutions are now aggressively committed to online learning bringing their strong academic reputations and large capital and personnel investments to the online learning space. Pre-contextual Grandbridge is watching its market share erode rapidly. The field has transformed while Pre-contextual Grandbridge has made only incremental changes in approach (e.g., adopting new electronic learning technologies).

Enrollments are down and Pre-contextual Grandbridge is facing an uncertain future as it enters a five-year strategic planning cycle. With drivers that potentially threaten the university's future viability, the executive leadership recognizes that Pre-contextual Grandbridge needs to take a major step forward and re-establish itself as leader in online learning and teaching. Terms like *transformational, world class, sector leading, innovative learning and teaching, and profound change* have become part of the executive lexicon as it prepares the community to enter the strategic planning cycle.

Pre-contextual Grandbridge's President Douglas Harper has convened a meeting of the university's leadership including the vice-presidents, directors and deans as well as faculty and student representation to initiate a strategic planning process that will focus first on learning and teaching. The leadership group begins its work by revisiting Pre-contextual Grandbridge's mission statement: *"To pursue excellence in learning and teaching by providing students with a world class experience in a distinctive learning environment. Grandbridge serves the state and nation by producing responsible well-educated graduates capable of advancing knowledge and providing responsible leadership."* The strategic planning group uses the statement as a term of reference to identify a number of strategic priorities. They include:

- To deliver a world-class online learning experience;
- To maximize the learning outcomes for students by improving student retention and graduation rates;
- To become the national leader in online learning by supporting innovative practice;
- To create a culture of innovative teaching excellence;
- To support the best in contemporary learning technologies and learning spaces[1];
- To become a national leader in quality assessment in higher education.

President Harper believes there are clear distinctions to be made between the strategic and the operational dimensions of the plan. He believes his role and that of the executive is to set the strategic agenda that will be enacted by the operational levels of management across the university. He asks the group to identify action steps, the parties responsible for implementation, and key performance indicators. The group undertakes to workshop each of the priorities. The Offices of Admissions and Student Affairs assume responsibility for the graduation and retention goals. Their efforts focus on expanding *Student Success at Grandbridge,* a support program for mentoring new and continuing students, strengthening the enrollment profile, and making more effective use of learning analytics data for predictive modeling. The latter includes progress and graduation rates as performance indicators. The Center for Excellence in Learning and Teaching (CELT) assumes responsibility for the three goals associated with innovation in teaching, technology, and online learning. The action plan includes creating four new *Quality Consultant* positions to support faculty in developing excellent online learning materials and to assume responsibility for the implementation of new quality assurance and assessment standards for online courseware. Two new positions are created to deliver professional capacity building for faculty members in best practice teaching and assessment. It is expected that improved faculty capacity will result in increased applications for promotion using the university's promotion framework for learning and teaching. The university's excellence in teaching and teaching innovation grant programs are expanded with a doubling of funds. Funding is also provided for the purchase of a new learning management system with expanded learning analytics

---

[1]The goals described here are not idiosyncratic to the example. They reflect the form and content of statements described in many university strategic plans.

capability. The Vice President (Academic) (VPA) is given the responsibility to create pathways between CELT and the faculties for professional capacity building and to ensure greater use of the promotion framework.

The leadership group identifies six major areas for measuring performance related to the learning and teaching goals. They are: to move Pre-contextual Grandbridge from the fourth to the second quartile in national rankings of teaching excellence on the National Assessment of University Teaching (NAUT); to achieve a 10% improvement in scores on the Grandbridge University Student Satisfaction Survey (GUSS); a 20% improvement in participation in the university's teaching award and grant programs; a 30% improvement in the number of faculty members that participate in the Center's faculty capacity building program and for 10% of faculty to have achieved promotion using the university's new promotion framework for learning and teaching. Further, the leadership sets a goal to establish a whole-of-organization review for the purpose of achieving what President Harper describes as greater organizational efficiencies in learning and teaching. This involves building a common approach to the management and delivery of online learning in ways that reduce idiosyncratic approaches and the duplication of effort within and across the faculties of the university. The Chief Financial Officer (CFO) is designated to lead the efficiency review. The leadership group sets the goal of achieving a 5% reduction in the costs associated with the design and delivery of Pre-contextual Grandbridge's online learning program. The planning group unanimously approves all of the strategic planning goals, performance indicators and the efficiency review. The group is confident that the plan can secure Pre-contextual Grandbridge's future.

Like most of its peer institutions, Pre-contextual Grandbridge has lofty aspirations reflected in its mission statement and strategic planning priorities. However, the community's limited understanding of its professional context means its priorities lack the kind of clarity and measurability required for effective strategic planning. Terms like *world class*, *national leader, maximize learning outcomes*, and *teaching excellence* do not possess a visible and comparable foundation in known professional practice at Pre-contextual Grandbridge or anywhere else and as a consequence have little operational value. If we were to ask the members of the strategic planning group to define these terms it is likely that there would be as many different responses as there are group members. The same would apply if members of the broader community were also asked to define the terms. The language of lofty aspiration functions as a substitute for the kind of specific goals that could be set if Pre-contextual Grandbridge possessed a genuine understanding of its context and a model of learning and teaching that exerted an attributable effect on the learning of its students. More disconcerting is the knowledge that Pre-contextual Grandbridge is not constituted with the systems, methods, and tools required to act upon these goals at scale.

The Pre-contextual Grandbridge approach to learning and teaching is an example of the effort chain described in Chap. 1—a loose coupling (Weick 1976) of awards, grants, governance, capacity building, promotion frameworks, and surveys none of which are underpinned by professionally controlled practice at a whole-of-organization scale. In line with Conway's Law, Pre-contextual Grandbridge's strategic planning mirrors its effort chain schema and design and the agency of those who work within its social structure. Its strategic plan amplifies its existing approach with more capacity building, more awards, more grant funding, and more promotion applications based on ambiguous standards and expectations.

Pre-contextual Grandbridge will purchase a new Learning Management System (LMS)—a commercial product already deployed by most of its competitors. The analytic capability of the LMS includes the measurement of log-ons, downloads, and navigation patterns, none of which have been shown to exert a proximal influence on learning and teaching (Bain and Drengenberg 2016). There is little in the way of research and experience to suggest that the Pre-contextual Grandbridge community will be able to extract any distinctive learning and teaching advantage from a system that is designed to meet the needs of effort chain universities.[2] An expanded CELT will develop its range of capacity building, awards, and grant programs increasing the scope of its exemplary approach to institutional improvement.

Pre-contextual Grandbridge's limited understanding of its learning and teaching context is also reflected in the way it compartmentalizes responsibility for its strategic planning initiatives. The strategic planning group assumes responsibility for the big picture planning and designates offices, roles and specific individuals to realize the plan's learning and teaching aspirations. The Chief Financial Officer leads the efficiency review. The Admissions and Student Affairs Offices assume responsibility for progress and retention. The CELT creates new positions to expand its capacity within and along the effort chain. Quality consultants assume responsibility for new quality standards and their expression in online courseware. The VPA is assigned the responsibility for building connections between the Center for Excellence in Learning and Teaching and the faculties. The leadership group believes that building these connections is an appropriate role for the VPA although how this goal will be accomplished is unclear. There are no pathways or networks within the compartmentalized and hierarchical organizational structure at Pre-contextual Grandbridge to make these connections substantive because the responsibility for learning and teaching is sequestered in centers and offices. The effect of compartmentalizing is to make the offices and designated individuals on the Pre-contextual Grandbridge organizational chart the locus of responsibility for learning and teaching instead of the whole community. There is no model that would make possible the systemic understanding of learning and teaching required to disperse control and distribute leadership and responsibility. Entities and positions on the organizational chart serve as proxies for a whole-of-organization professionally controlled approach and the overall lack of understanding of learning and teaching.

Without a proximal understanding of the learning context, Pre-contextual Grandbridge also defers to distal measures of performance that are unlikely to result

[2]This idea is discussed extensively in Bain and Drengenberg (2016) where research in the compulsory and higher education sectors are employed to show how the lack of professional control in higher education severely limits the role and application of technology and specifically learning management systems with respect to their influence on learning and teaching. This idea is also taken up in Chap. 7.

in genuine improvement in student learning. It counts the participation in effort-chain programs that have an unsubstantiated effect on learning and teaching and employs survey measures of questionable validity to determine learning and teaching effect. Pre-contextual Grandbridge is doing more of the things that resulted in its current performance and produced its precarious circumstances. There is no form or methodology for executing the plan beyond the existing structures that are most likely to produce more of the same.

The efficiency review also mirrors its effort-chain compartmentalized schema. It focuses on making cost saving changes in roles and offices because it is through these roles and offices that it defines and understands learning and teaching. Reducing cost is a relevant outcome for the university given its precarious fiscal circumstances. However, the efficiency review is unlikely to improve the quality of learning and teaching because Pre-contextual Grandbridge does not have a clear professional understanding of learning and teaching quality and productivity and how they are served by the university's organizational structure and process. Its inability to understand quality and productivity means it does not really know where to best spend or save money and the impact of those decisions. Its effort is analogous to a hospital reviewing the efficiency of its organization and processes without a professional understanding of the medical practice or patient outcomes it delivers. We noted in Chap. 1 the inability to effectively link the organizational structure of universities to the quality of learning and teaching practice (e.g., Knapp and Brandon 1998; Stensaker 2008; Zundans-Fraser and Bain 2015). A number of authors have pointed to a counter productive relationship between university organization and learning and teaching practice (e.g., Coate and Tooher 2010; Knapp and Brandon 1998; Zundan-Fraser 2014). Pre-contextual Grandbridge may meet its immediate strategic planning goal by achieving organizational efficiencies, although, any short to medium term fiscal advantage is unlikely to alter Grandbridge's trajectory because its cost savings are at best only distally connected to the quality, productivity, and distinctiveness of its learning and teaching approach.

According to Conway's Law, the ways Pre-contextual Grandbridge is seeking to meet its strategic priorities makes complete sense—the existing organizational schema and structure is mirrored and then amplified in the strategic planning process, actions, and outcomes. It is here that we see Pre-contextual Grandbridge's problem most clearly. Its effort-chain approach characterized by a poor understanding of context (Bain and Drengenberg 2016) and lack of professional control (Bowker and Star 2000) drives and defines its schema for strategic planning. The planning group sets goals, determines actions, and performance indicators that will, with all likelihood be accomplished successfully in an effort-chain sense. The university's leadership by meeting its immediate strategic planning goals will claim a victory and market its success to the community. The effort chain schema frames the effort chain agency of those who work within the structure. Pre-contextual Grandbridge will be seen to be proactive, responsible, and purposefully engaged in securing its future although little will change in terms of improving the quality of learning and teaching at scale or addressing the university's precarious position in the higher education market place.

2.3 The Futility of "Industry Standard"

Universities organized for the effort chain approach cannot realize their whole-of-organization strategic objectives for learning and teaching improvement at scale. They are designed for the elective, selective approach described in Chap. 1. As we described, factors outside of the university's sphere of learning and teaching influence continue to account for most of the variance in student growth (Pascarella and Terenzini 2005). The strategic planning effort at Pre-contextual Grandbridge described in Example 2.1 also highlights the way strategic planning can initiate whole-of-organizational changes (e.g., the efficiency review) that are unlikely to alter the quality and productivity of learning and teaching. Administrative restructuring efforts and efficiency reviews are common in higher education. They create the appearance of meaningful change about learning and teaching needs although they are a weak proxy when they do not connect change in an organization's design or administrative structure to a deep understanding of its learning and teaching context.

The greatest single challenge associated with the improvement of learning and teaching in higher education is recognizing that the current effort chain approach and the architectural structures that serve it are incompatible with the whole-of-organization improvement in learning and teaching represented in mission and value statements and strategic plans and so desired by universities. According to Bird (2012), architecture needs to reflect purpose. In the current context this means recognizing and then doing something about the inconvenient truth described in Chap. 1—the aspirations, productivity, distinctiveness, transformational change and student outcomes are not possible employing an effort chain approach.

Generations of leaders have claimed or aspired to learning and teaching excellence in a manner akin to the Pre-contextual Grandbridge strategic planning approach. Most plans are never subjected to the kind of genuine empirical scrutiny required to determine improvements in quality and productivity at scale. The performance needle may move in terms of the attainment of specific planning goals (e.g., more promotions, more award winners etc.), although as we have seen from the research on student outcomes of higher education not much changes. The central question for this chapter is how to break out of the traditional strategic plan/effort chain cycle and replace it with an approach capable of the systemic influence aspired to by so many universities.

The first step in doing so is to recognize that there can be no systemic effect unless universities establish systemic professional control (Bowker and Star 2000) over the learning and teaching context. To do so requires a model of learning and teaching based on comparable and visible professional practice that makes an attributable connection between learning and teaching at a whole-of-organization level. As we saw in Chap. 1, when standards and priorities are underpinned by a professionally controlled model of practice the issue of whether the syntax and semantics of the language used to express goals or standards are general or specific fall away because the statements irrespective of language are underpinned by a shared professional understanding of what learning and teaching means (e.g., RACGP 2012).

2.4 Commitments

The cornerstone of professional control in a self-organizing university is a commitment to the research and practice of learning and teaching—those things that have been shown to influence student achievement and related learning outcomes. The schema for a self-organizing university is built upon a conceptual and practical commitment to those things that exert a proximal influence on quality learning and teaching. As such the first step in the SOU process is to identify those things that the university believes will make a difference in learning and teaching.

The idea of developing commitments in the SOU comes from research on self-organizing systems that shows highly complex natural and biological systems including human endeavors are often based upon a set of very simple rules (Pascale et al. 2001). The highly successful internet company EBay provides a perfect example of simple rules driving a human system. Simple rules that establish the reputation of sellers, secure the reliability of payment combined with constant feedback and the mediation of internet technology facilitate a global exchange of goods without inventory, storage management and control, logistics systems, and the extensive capital infrastructure required in a traditional model for the sale of goods. EBay provides a self-organizing platform for the agency of individual actors interested in buying and selling.

Identifying commitments need not be an act of invention. As we noted in Chap. 1, and in our work on the research-base for the SOU (Bain and Zundans-Fraser 2016), developing commitments using the SOU approach involves an examination of the extensive literature on models of educational productivity and the broader educational productivity research to build the cornerstones of a distinctive learning and teaching model.

2.5 Crowdsourcing Commitments

One of the challenges associated with traditional strategic planning is that it suffers from the compartmentalizing we saw in evidence in Example 2.1. Pre-contextual Grandbridge "in"sources the process of planning the future of the university to a selected leadership group. The group is then faced with the challenge of "out" selling the product of its efforts back to a broader community that may in part feel disenfranchised by its lack of participation. In many cases insourcing initiates a continuum of response from benign indifference to a counteractive effort by those who feel left out of the process. The community is positioned as consumers of the plan by virtue of not being involved in its creation. We saw this in Example 2.1 where the president needed to market or pitch the new Pre-contextual Grandbridge strategic plan back to the community.

An alternative way to develop commitments is to invite broad-based participation by the whole community from the outset by using principles of crowdsourcing to identify what the community believes about the learning and teaching foundation of the university's model. Instead of design by representation, the design process can become an emergent expression of the agency of the whole community. Crowdsourcing is the process of outsourcing a task to an undefined (and generally large) network of people in the form of an open call (Howe 2006; TechTarget 2015).

The earliest example of crowdsourcing occurred in 1858 when a group of over 800 volunteer readers were contracted to collect words from all available books for the Oxford English Dictionary (TechTarget 2015). Aitamurto and Landemore (2015) identify three forms of crowdsourcing: ideation for broad-based idea generation; argumentation for the review; and evaluation of courses of action and micro-tasking where crowdsourcing is employed to outsource specific parts of a process to large groups of people. The SOU employs all three forms of crowdsourcing in its phases as part of an emergent community-based change methodology. In the case of developing commitments the crowdsourcing methodology focuses on ideation. The goal is to bring the community into a generative conversation about what it believes should be the key elements of the universities approach to teaching and learning.

Orientation involves providing detailed information about the focus or content area for the crowdsourcing call or task. Ideation is the input provided by the crowd. Evaluation is the way in which value is assigned to the information (e.g., voting processes). In Example 2.2 we unpack Fig. 2.1 and show how Grandbridge took a different approach to address its circumstances using the SOU to build the foundations for a learning and teaching model.

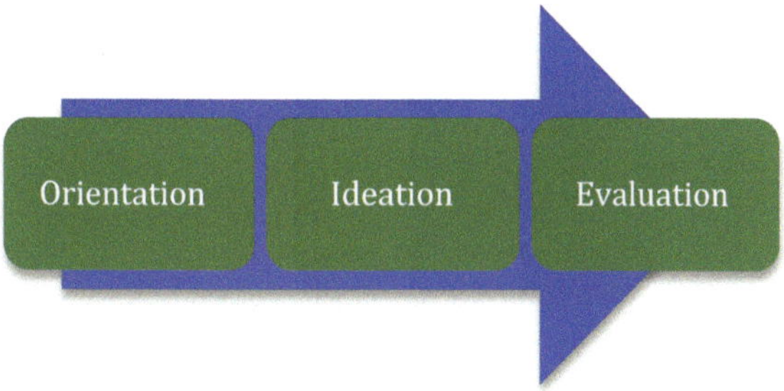

Fig. 2.1 Phases of crowdsourcing in the SOU (Aitamurto and Landemore 2015)

Example 2.2 Grandbridge SOU Crowdsources its Learning and Teaching Commitments

President Harper has come to three important conclusions about the future of the university. First, it needs to make profound improvements in the approach taken to online learning and teaching to be competitive. Grandbridge SOU requires a new and genuinely distinctive approach that focuses on the quality of its courseware. Second, the effort chain approach taken to leading learning and teaching to date has not delivered the outcomes expected and required for the kind of profound transformational change required to secure the university's future. Third, the traditional top-down leadership-dominated approach to strategic planning is not adequate for the scope and magnitude of the change Grandbridge SOU requires.

As Grandbridge SOU enters what may be the most critical strategic planning cycle in its history the president decides on an approach to build a new model of learning and teaching that will involve the whole community from the outset. He knows that the kind of change required at Grandbridge SOU will be disruptive and controversial making it essential to invite broad and informed participation.

To orient the community, he asks the Director of the Centre for Excellence in Learning and Teaching to develop a video about the future of learning and teaching in universities using leading research on models of learning, and excellence in online learning. He also asks for a concise user-friendly presentation on evidence-based learning and teaching approaches from the educational productivity research. The Centre also develops some short and accessible documents describing case studies of evidence-based practice in action in program and course design and in the design and implementation of learning experiences.

The president directs his office to create a website entitled Grandbridge SOU that will host the orienting materials. The resources of the website are also linked to a large number of faculty and student pages on the Grandbridge SOU website including links to alumni. The VPA asks the deans and department heads to develop presentations for faculty meetings on the change process using the orientation materials on the SOU website. He calls upon the leadership at all levels to build an advanced understanding of the educational productivity and excellence in online learning research in readiness for leading a change process across the university. He encourages these leaders to ask for further clarification and support if required. It is made clear that he expects these staff to actively lead the process within their own faculties and departments.

Finally, the president develops a video introduction to the change process. He describes the past success and current challenges frankly, indicating the drivers for change and the critical issues facing the university. He explains the phases of the process from developing commitments through to building out a new organizational design for learning and teaching that includes new technologies, feedback approaches, human resource requirements and curriculum. He invites the community including alumni to be involved in all phases and makes a personal commitment to a comprehensive and participatory process. The presentation concludes by introducing an immediate call to action by crowdsourcing the university's learning and teaching commitments. He invites the community to identify the approaches to learning and teaching they believe should be the foundation of the Grandbridge SOU approach. The VPA follows up with presentations to all divisions and offices that echo the president's message.

In the second ideation phase crowdsourcing software[3] is used to collate and map responses submitted by email, through the universities social media channels, and recorded directly in

[3]There are a number of commercially available collaborative software platforms that can be employed for compiling sharing and analyzing crowdsourced data.

the software. The community is encouraged to deliberate on the submissions, again using crowdsourcing software that encouraged argumentation and justification of the commitments. A list of commitments is returned to the community for a third phase evaluative voting process where participants are asked to vote their top 10 commitments. The responses are further analyzed and evaluated by a group of internal and external experts that provide feedback to the community. The evaluation process involves cross-checking the input of the community with the orienting information to identify continuities and discrepancies with the best evidence in the sector. This process is undertaken by the staff of CELT. Ten key commitments are identified. These are:

1. *Collaborative Design and Decision-making*;
2. *Universal Design for Learning* Principles applied at Program and Course Level;
3. Student *Mastery Focus* for Learning Design and Delivery;
4. *Active Lecture* Design;
5. *Learning Strategies* for Students;
6. *Reflection* based upon *Multiple Feedback* Sources (not just student evaluations);
7. *Criterion-based Assessment*;
8. *Best-evidence Teaching Approaches*;
9. *Learning and Teaching Analytics*;
10. *Curriculum that is mapped to standards and constructively aligned.*

Harper knows the commitments represent a major step toward professional control for Grandbridge SOU. The university is setting a baseline for the future of learning and teaching using practices that are known to exert a demonstrable and powerful effect on student learning. The leadership has built an understanding of those practices with the community as opposed to doing so on its behalf.

There are a number of big ideas about the SOU represented in the example. First, president Harper pursues the wisdom of the crowd[4] as it shapes the university's future. Harper, by initiating the change process with a crowdsourcing approach is modeling the way he hopes learning and teaching at Grandbridge SOU will work in the future. The SOU is based on dispersed control, the agency of those directly involved with learning and teaching, and their emergent feedback. Harper intends to build Grandbridge SOU's learning and teaching approach using the same principles of self-organization that he hopes will underpin its day-to-day operation.

Second, the commitments have the potential to be much more than a list of undefined aspirations. Each item on the list is associated with a visible and comparable educational practice that has clear and definable implications for the day-to-day work of the university. As we will describe in the chapters that follow, the practices listed become the cornerstones of a new schema that includes the way the university is organized and governed; the way feedback happens and the way the university builds capacity. The practices also provide valid and reliable meaning to evaluation, promotion frameworks and shapes Grandbridge SOU's ongoing agenda for innovation. At this point potential is definitely the operative consideration. Much work needs to be done if the commitments are to realize their full potential in the SOU approach.

[4]A reference to the book by James Suroweicki about the way decisions sourced by crowds are better than individual experts.

A third important takeaway from the example pertains to the way the university's middle management is engaged from the outset. The new approach must be led assertively and actively. As such, he needs to craft a different role for leadership. Instead of *being the solution* in the compartmentalized effort chain approach, leadership assumes a responsibility for the process, stimulating interest in the SOU crowdsourcing approach by promoting opportunities for input and sustaining the enthusiasm necessary to ensure that the Grandbridge SOU community benefits from its collective intelligence. President Harper directs his VPA to engage the university's middle managers actively in the orientation process.

Harper also knows that most of the educational leaders at Grandbridge SOU do not know a lot about professional education; while they may be deans of faculties and associate deans for learning and teaching, the research and practice of best-evidence quality teaching, learning and curriculum development is not in the experience base of many. In an effort chain approach to learning and teaching compartmentalization precludes the need for the community as a whole to have a professional knowledge of learning and teaching practice. Administering learning and teaching takes priority over a deep understanding of practice. Grandbridge SOU cannot get better at learning and teaching without knowing more about what quality learning and teaching means.

Harper knows that if his leadership does not build capacity and as a result comes to fear the change it will likely passively or actively resist it. He wants to ensure that his leadership team is working in advance of the expectations of the change process by building capacity with the substance of the change they will lead. Harper has worked out that the dire situation at Pre-contextual Grandbridge precludes a superficial "claim of victory" type change approach and that a credible approach means transforming learning and teaching in ways that will be both substantive and distinctive. This means transforming the learning and teaching capacity of the whole community. To do so he must have his leadership on board.

Fourth, while the president is pursuing a model of distributed leadership for the future, he is also nailing his colors to the mast of Grandbridge SOU as he initiates and leads the change process. We saw in Example 2.1 how the traditional strategic planning approach sequesters and limits the role of the executive to high-level strategic matters including building the strategic plan. A change as big as Grandbridge SOU blurs the distinction between strategy and operations. Everyone at all levels of the organization will have to know and do things differently. Harper knows that he cannot change all of Pre-contextual Grandbridge with a compartmentalized approach that creates hard distinctions between strategy and operation and where his involvement is limited to mentioning Grandbridge SOU in speeches and memoranda or directing his managers to implement a strategy. He is leading a learning and teaching organization that does not know a lot about the professional practice of learning or teaching. To lead, *he* will need to learn more and demonstrate to the broader community his willingness to be a learner. The community needs to see and sense his commitment to the change process and that of his leadership team if it is to take the effort seriously.

Harper also knows creating a self-organizing university is not autogenetic. Even if the university he is seeking to help create will be one driven by the bottom-up emergent feedback, getting that model in place is an act of design that requires leadership. By linking the new SOU website to his office, taking an active role in the orientation component of the crowdsourcing model, and actively leading the whole process, Harper is beginning to walk the talk of self-organizing transformational change by providing the order in the design process required to give up control (Bain 2007). He recognizes that there is no fail-safe complete technology and absolute step-by-step instrumental process for the nuances of undertaking transformational organizational change. He will need to address questions and challenges for which solutions will not be immediately available and where his personal resolve and the commitment and skill of his leadership team will be necessary to move Grandbridge SOU forward.

2.6 From Commitments to Policy Using Commons-Based Peer Production

As we noted in the introduction to Example 2.1, one of the challenges faced by Pre-contextual Grandbridge in its traditional strategic planning mode is lending meaning to terms like *world class* and *innovative teaching* described in its strategic plan. These terms have little utility unless the community understands what they mean and can act upon them. By adopting the SOU approach, Grandbridge is seeking to avoid developing benign indistinctive statements of mission and values that will appear on websites and in marketing materials but have little practical day-to-day meaning for the community.

However, adopting a more professional lexicon is only a step forward if the community also knows and understands what those terms mean and can act upon them. The lack of pre-existing visible and comparable practice at Pre-contextual Grandbridge means knowledge of the professional language of education cannot be assumed to be the case any more than its understanding of terms like *innovative* and *world class* associated with its traditional strategic planning approach. Professional terms like *collaboration, learning strategies,* and *cooperation* may have quite explicit meaning in the educational literature although the research to practice gap in education means that such knowledge is not widely known or adopted (Grima-Farrell 2013; Hennessey and Dionigi 2013; Jenkins et al. 2003).

Grandbridge through its crowdsourcing approach has identified ten learning and teaching commitments, those things its community believes should be present in what Grandbridge does in all of its learning and teaching work. President Harper's intent is to avoid the commitments turning into educational versions of the lofty aspirational statements associated with the traditional strategic planning approach by ensuring that the community has an understanding of what they mean. To do so, the leadership team led by the VPA extends the crowdsourcing methodology to develop a learning and teaching policy. The policy is a first step toward making the commitments a practical driver for the normal learning and teaching work of faculty members and students.

Grandbridge SOU employs a process known as Commons-Based Peer Production (Benkler 2002) an approach used to develop open source software, industry standards and guidelines, and nanotechnologies (Maurer 2010b). Commons-Based Peer Production (CBPP) can be defined as any process whereby individuals can freely and openly contribute to a common pool (of knowledge, code, and design) (Bauwens 2014). CBPP is a decentralized alternative to hierarchical development processes for the peer production of what are frequently complex products (Benkler 2016). CBPP pools collective intelligence in a proactive production process that permits the inclusion of a broad base of perspectives and motivations (Vuculescu 2012). Wikipedia is the best known and most widely used example CBPP where anyone can author or edit a contribution within a simple rule and oversight structure. There are three phases in the CPBB for policy development in the SOU. First, the community builds definitions and examples to clarify the meaning of its commitments. The instantiated commitments are then used to develop policy statements. The statements are then evaluated, refined, and organized to become a policy document. Example 2.3 describes the process as employed by Grandbridge.

Example 2.3 Building a Learning and Teaching Policy at Grandbridge

The Grandbridge SOU leadership knows that getting to scale with a university-wide model of learning and teaching means building whole-of-organization commitment from the ground up. President Harper invites all offices, faculties and divisions of the university as well as all individuals to:

1. Revisit, refresh and familiarize themselves with the orientation information on the SOU website;
2. Submit definitions of the key terms- any or all;
3. Provide practical examples of the key terms/concepts in action in the design of a course, its delivery or in the way students engage with the commitment.

The process places no limits on the number of submissions by any single individual or office, or the extent of their contributions. A contributor can define one or all commitments and submit one or many examples. The instantiation task is framed by two rules. The definition:

1. Must be supported by verifiable empirical research[5];
2. Must refer to practice in the design, delivery or student engagement with learning and teaching practice.

[5]This rule was included recognizing the contestation that exists among educators about what constitutes evidence and empirical research support. The intent was to surface multiple perspectives and required respondents to make a case for their definitions.

> Universal Design for Learning means a scientifically derived framework for guiding educational practice that:
> 1. Provides flexibility in the ways information is presented, in the ways students respond or demonstrate knowledge and skills, and in the ways students are engaged; and
> 2. Reduces barriers in instruction, provides appropriate accommodations, supports, and challenges, and maintains high achievement expectations for all students including those with a disability. NCUDL (2016).

Fig. 2.2 Example of a commitment definition for universal design for learning

Examples are also included on the SOU website. Figure 2.2 describes an example of a definition for Grandbridge SOU's second commitment provided by the office of student services.

As with the initial crowdsourcing, the request for participation is made to alumni, the university governance, and all academic and non-academic entities, student groups, and to every individual in the community. Managers, Deans and Heads of Department are asked to find time in the regular meeting schedule for the CPPP process. Members of the community are asked to respond through their affiliated office/entity, as individuals and either anonymously or name identified. Participants submit new content related to the commitments, or edited existing submissions. The CBPP software records the original submissions, the edit trail and the network of action around each of the commitments. The editing option produces a tighter set of definitions and examples as contributors when seeing exemplars that were close to their own intended input chose to refine and/or develop existing definitions and/or exemplars over adding new ones. The initial submission rules are used to review, refine and consolidate submissions to produce a statement, definition and set of examples for each commitment. Figure 2.3 provides an excerpt of the instantiated Grandbridge commitments after the CBPP phase.

In the second phase of the Commons-based approach the matrix of commitments, their definition and examples in the format described in Fig. 2.3 are posted to the SOU website. The community is asked to contribute statements about how the commitments should be expressed as policy. Again, community members contribute through the entity with which they are affiliated at Grandbridge SOU or independently. Contributors also have the option of contributing original statements or editing existing ones. A period of one month is allocated for developing the statements. Figure 2.4 describes two policy statements for Learning Analytics contributed by members of the Information Technology Division at Grandbridge SOU. The IT group employed the best practice literature in the learning analytics field to craft the draft statements and included citations to the sources they employed.

In the third phase of the CPBB process, the pool of completed statements is organized according to three categories related to expectations of the university, of students and of faculty. All of the statements are posted on the SOU website and the community is asked to vote with a simple like/dislike voting system. In a final step the community is asked to identify any areas gaps or omissions before the statements are compiled into a policy document. A draft of the policy is then posted on the SOU website for reaction and comment. In this case the community is not asked to actively edit the statements but to make comments and suggest final edits/changes. Figure 2.5 shows a list of example statements from Grandbridge.

SOU's learning and teaching policy. Each of the statements is accompanied by an active *example link* whereby the examples derived from the first phase of the CBPP process are linked to the policy enabling users to see what every policy statement means in action.

Grandbridge SOU requires a change methodology commensurate with the magnitude of its challenging circumstances. President Harper knows that by discarding the effort chain approach he will invoke profound disruption for faculty and students. He believes that such a change cannot be *done to* the community and it must be undertaken in a manner that creates the conditions for total involvement. He also knows the kind of change he is embarking upon is rare and subject to high levels of failure in large organizations. Harper also understands the barriers, the

Commitment	Definition	Example
Best-evidence teaching approaches	The use of teaching approaches that exert a visible effect on student learning based on empirically verifiable learning features (Hattie 2008).	The design and delivery of Cooperative Learning includes activities that require mutual interdependence and individual accountability (Slavin, 1996; Springer, Stanne & Donovan, 1999). The design of lectures should be active including opportunities for guided and independent practice of key ideas and concepts.
Collaborative Decision-making	An agreed process of interaction by a group of individuals working together for the best resolution to a common problem or goal.	Meetings employ a collaborative problem-solving process that includes problem definition, solution generation action-planning and evaluation (Friend & Cook, 2014)

Fig. 2.3 Defined commitment, definition and example for a best-evidence teaching approach and collaborative decision-making

> Learning and Teaching Analytics at Grandbridge SOU involve gathering and using data about Grandbridge's commitments—those things that the university believes produce quality learning and teaching.

> Learning and Teaching Analytics data at Grandbridge SOU should be gathered using valid and reliable practices of the highest quality to provide the community with transparent and continuous access to data and reportage employing the highest ethical standards (Sclater, 2014; Willis, 2014).

Fig. 2.4 Policy statement examples

> Grandbridge SOU students master program and course content in rigorous criterion-based assessment tasks mapped to sector and professional standards. Examples
>
> Feedback about learning and teaching at Grandbridge SOU is continuous and derived from valid and reliable measurement of programs, courses, faculty and student performance related to the university's commitments. Examples
>
> All programs and courses at Grandbridge SOU are designed and delivered employing best evidence teaching approaches including active learning design, cooperative learning, and problem-based learning, and learning strategies. Examples.
>
> All Grandbridge SOU programs and courses are mapped to sector and professional standards at the learning experience and assessment task levels. Examples

Fig. 2.5 Example learning and teaching Policy Statements

change fatigue, the expertise gap and the fear and cynicism that frequently accompany large scale change efforts (Maurer 2010a). The history of change in education would suggest that despite his commitment to an inclusive approach, that few change efforts meet with the success envisaged by their advocates. It is under this tension that he must negotiate a path. He is also aware that there is a huge diversity of perspectives on learning and teaching at Grandbridge SOU and that crowdsourcing and CBPP are not panaceas. Only a percentage of the community members will avail themselves of the crowdsourcing and CBPP opportunities for building commitments and policy. The process needs to be highly transparent where choices are justified, entities within the organization need to be engaged and respected, and where there can be no sense of an end-run around the approach (Landemore, 2012, 2014). Most important, President Harper has created an emergent opportunity footprint. He has invited everyone to be a part of the change and is following through on a broad-based participatory methodology that makes the opportunity to be involved real and accessible.

The SOU approach to developing commitments is a first step in developing a self-organizing university. Grandbridge SOU has replaced its amorphous strategic priorities with a set of commitments to professional practice and a policy that describes those commitments. There is no expectation that those commitments can become a day-to-day reality without profound change to the way the organization is constituted for learning and teaching. In the SOU, and in line with Conway's Law, there is also no expectation that a set of commitments will exert an autogenetic effect on an effort chain organization. For example it is unlikely that a learning management system can gather data about Grandbridge SOU's new commitments if it has been designed and developed to gather a different kind of data. As obvious as this may seem, it is a common trap for university's seeking transformational change. Grandbridge SOU needs to be designed for comparable and visible professionally controlled learning and teaching in order to make its commitments a reality at scale. To do so, means embedding the commitments and policy in the way the University functions, the next step in the SOU.

Finally, if there is a parallel in the pre-contextual university for the kind of process described in this chapter, it is most like the input and approbation sought by pre-contextual university leaders for efforts to develop mission and value statements. These efforts seek to identify and establish support for statements that distinguish the pre-contextual university from its peers. They pre-occupy leaders as they seek to build and message their distinctive value propositions. However, as we noted in Chap. 1 these efforts are more likely to produce rhetorical sameness than distinctiveness (Dvorak and Busteed 2015), bear little relationship to the quality of learning and teaching, and have become a time-consuming proxy activity for the real challenging work of bringing an understanding of quality to what universities do in the domain of learning and teaching.

Takeaways

The key takeaways from this chapter are:

- Terms like *world class*, *national leader, maximize learning outcomes*, and *teaching excellence* do not possess a visible and comparable foundation in known professional practice and as a consequence have little operational value in focusing a change process;
- The factor most likely to influence a strategic plan and its outcomes is the way an organization is constituted as it enters the planning process;
- Leaders who do not feel capable of leading change are susceptible to become passively or assertively resistant. They need to be an early focus of support and preparation in an inclusive approach to change;

- When goals and priorities are underpinned by professionally controlled practice, the syntax and semantics of the language included in planning documents and statements of belief fall away in significance;
- A university that seeks the benefits of self-organization needs to identify its simple rules or commitments. They are the educational processes and practices that exert a proximal influence on its core purpose or function—learning and teaching;
- Crowdsourcing and CBPP outsource a traditionally exclusive process to those who have most at stake in the way an organization changes and the end product of that change;
- An inclusive participatory approach to developing commitments sends a powerful message irrespective of the levels of participation in the process;
- Transformational commitments require a transformed organizational design;
- While the goal of the SOU is to create a self-organizing system, such a transformation is not autogenetic and requires active, assertive, and persistent leadership;
- Change methodology needs to reflect its intended purposes and outcomes.

References

Auhl, G. (in progress). The development of pre-service teachers' schema for teaching students with diverse learning needs.

Aitamurto, T., & Landemore, H. (2015). Five design principles for crowdsourced policymaking: Assessing the case of crowdsourced off-road traffic law in Finland. *Journal of Social Media for Organizations, 2*(1), 1–19. Retrieved from http://www2.mitre.org/public/jsmo/pdfs/02-01-5-design-principles.pdf.

Bain, A., & Zundans-Fraser, L. (2016). *Rising to the challenge of transforming higher education: Designing universities for learning and teaching*. New York: Springer Verlag.

Bain, A., & Drengenberg, N. (2016). *Transforming the measurement of learning and teaching in higher education*. New York: Routledge.

Bain, A. (2007). *The self-organizing school. Next generation comprehensive school reforms*. Lanham, MD: Rowman & Littlefield.

Bauwens, M. (2014). Commons based peer production: An introduction. Retrieved from https://www.boell.de/en/2014/07/08/commons-based-peer-production-introduction.

Benkler, Y. (2016). Peer production and cooperation, forthcoming in J. M. Bauer & M. Latzer (Eds.), *Handbook on the economics of the internet,* (pp. 81–119). Cheltenham and Northampton, Edward Elgar.

Benkler, Y. (2002). Coase's penguin, or Linux and the nature of the firm. *Yale Law Journal, 112,* 369–446.

Bertolini, M., Duncan, D., & Waldeck, A. (2015, December). Knowing when to reinvent. *Harvard Business Review*. Retrieved from https://hbr.org/2015/12/knowing-when-to-reinvent.

Bird, J. (2012). Should you care about Conway's law? Retrieved from http://swreflections.blogspot.com.au/2012/10/should-you-care-about-conways-law.html.

Bowker, G., & Star, S. (2000). *Classification and its consequences*. Cambridge, Massachusetts: The MIT Press.

CFAR. (2005). A summary of best practice approaches in strategic planning processes. Retrieved from http://trugroup.com/whitepapers/tru-strategic-planning-best-practice.pdf.

Coate, K., & Tooher, M. (2010). The Galway symposium on design for learning: Curriculum and assessment in higher education. *Teaching in Higher Education, 15*(3), 347–354. doi:10.1080/13562511003740924.

Conway, M. E. (1968). How do committees invent. *Datamation, 14*(5), 28–31.

Dvorak, N., & Busteed, B. (2015, August, 11). It's hard to differentiate one higher-ed brand from another. *Gallup Business Journal,* Retrieved from http://www.gallup.com/businessjournal/184538/hard-differentiate-one-higher-brand.aspx.

Grima-Farrell, C. (2013). Bridging the research to practice gap to respond to student and teacher needs in Catholic school settings: A collaborative school and university approach. *Journal of Catholic School Studies, 84*(2), 66–75.

Friend, M., & Cook, L. (2014). *Interactions: Collaboration skills for school professionals* (7th ed.). New York: Pearson.

Hanover Research. (2013). Strategic planning in higher education: Best practices and benchmarking. Retrieved from http://www.hanoverresearch.com/media/Strategic-Planning-in-Higher-Education-—-Best-Practices-and-Benchmarking.pdf.

Hattie, J. (2008). *Visible learning: A synthesis of over 800 meta-analyses relating to achievement.* New York: Routledge.

Hennessey, A., & Dionigi, R. A. (2013). Implementing cooperative learning in Australian primary schools: Generalist teachers' perspectives. *Issues in Educational Research, 23*(1), 52–68.

Howe, J. (2006). *The rise of crowdsourcing. Wired Magazine, 14*(6), 1–4.

Jenkins, J., Antil, L., Wayne, S., & Vadasy, P. (2003). How cooperative learning works for special education and remedial studies. *Exceptional Children, 69*(3), 279–292.

Knapp, M. S., & Brandon, R. N. (1998). Building collaborative programs in universities. .In M. S. Knapp (Ed.), *Paths to partnership: University and community as learners in inter professional education.* Oxford: Rowman & Littlefield Publishers.

Landemore, H. (2014). We, all of the people: Five lessons from Iceland's failed experiment in creating a crowdsourced constitution. *Slate, July 13.* Retrieved from http://www.slate.com/articles/technology/future_tense/2014/07/five_lessons_from_iceland_s_failed_crowdsourced_constitution_experiment.2.html.

Landemore, H. (2012). Why the many are smarter than the few and why it matters. *Journal of Public Deliberation, 8*(1), Retrieved from http://www.publicdeliberation.net/jpd/vol8/iss1/art7.

MacCormack, A., Baldwin, C., & Rusnak, J. (2012). Exploring the duality between product and organizational architectures: A test of the "mirroring" hypothesis. *Research Policy, 41*(8), 1309–1324.

Martin, R. (2014, January–February). The big lie of strategic planning. *Harvard Business Review.* Retrieved from https://hbr.org/2014/01/the-big-lie-of-strategic-planning.

Maurer, R. (2010a). Why 70% of change fails. Retrieved from http://www.reply-mc.com/2010/09/19/why-70-of-changes-fail-by-rick-maurer/.

Maurer, S. (2010b). Five easy pieces: Case studies of entrepreneurs who organized private communities for a public purpose. Retrieved from https://gspp.berkeley.edu/assets/uploads/research/pdf/ssrn-id1713329.pdf.

NCUDL. (2016). UDL provisions in the higher education opportunity act (P.L. 110–315). Retrieved from http://www.udlcenter.org/advocacy/referencestoUDL/HEOA.

Pascale, R. T., Millemann, M., & Gioja, L. (2001). *The new laws of nature and the new laws of business.* New York: Crown Publishing Group

Pascarella, E., & Terenzini, P. (2005). *How college affects students (Volume 2): A third decade of research.* San Francisco: Jossey-Bass.

RACGP. (2012). *Standards for general practice* (4th ed.). Retrieved from http://www.racgp.org.au/your-practice/standards/standards4thedition/.

Sclater, N. (2014). Code of practice for learning analytics: A literature review of ethical and legal issues. Retrieved from: http://repository.jisc.ac.uk/5661/1/Learning_Analytics_A-_Literature_Review.pdf.

Slavin, R. E. (1996). Research on cooperative learning and achievement: What we know, what we need to know. *Contemporary Educational Psychology, 21,* 43–69.

Springer, L., Stanne, M. E., & Donovan, S. S. (1999). Effects of small-group learning on undergraduates in science, mathematics, engineering and technology: A meta-analysis. *Review of Educational Research, 69*(1), 21–51. doi:10.3102/00346543069001021.

Stensaker, B. (2008). Outcomes of quality assurance: A discussion of knowledge. *Methodology and Validity Quality in Higher Education, 14*(1), 1470–1081.

Suroweicki, J. (2004). *The wisdom of crowds*. New York: Anchor Books.

TechTarget. (2015). Crowdsourcing defined. Retrieved from http://searchcio.techtarget.com/definition/crowdsourcing.

Vuculescu, O. (2012). Commons based peer production for problem solving – perspectives from evolutionary computation. Retrieved from http://pure.au.dk/portal-asb-student/files/45294059/commons_based_pp.pdf.

Waldrop, M. M. (1992). *Complexity: The emerging science at the edge of order and chaos*. New York: Touchstone, Simon and Schuster.

Weick, K. E. (1976). Educational organizations as loosely coupled systems. *Administrative Science Quarterly, 21,* 1–21.

Willis, J. (2014). Learning analytics and ethics: A framework beyond utilitarianism. *Educause Review*. Retrieved from https://net.educause.edu/ir/library/pdf/erm0742.pdf.

Zundans-Fraser, L., & Bain, A. (2015). How do institutional practices for course design and review address areas of need in higher education. *Higher Education Research and Development, 35*(4), 841–853. doi:10.1080/07294360.2015.1137883.

Zundans-Fraser, L. (2014). *Self-organisation in course design: A collaborative, theory-based approach to course development in inclusive education*. (Unpublished doctoral dissertation). Charles Sturt University, Bathurst.

Chapter 3
Designing Universities for Better Learning and Teaching

3.1 Introduction and Overview

In this chapter, we explain the cornerstone of the self-organizing university (SOU) approach—embedded design. Embedded design translates a university's big ideas about learning and teaching into day-to-day normal practice. We build a rationale for embedded design and show how the principle is applied to the organizational design elements of the SOU to create a university capable of professionally controlled practice at scale. We use three examples that extend the crowdsourcing and commons-based peer production process described in Chap. 2; the first in the area of technology; the second focusing on student feedback; and the third, for developing a promotion framework. We also describe an agent-based software development approach for the design of technology tools. Our goal is to explain the embedded design principle, its importance, and show how it works. Subsequent chapters on emergent feedback, governance, technology, and agency take up in detail the elements discussed here and further develop what embedded design means and looks like in a self-organizing university.

Before defining and describing embedded design, it is important to first consider why this pivotal element of the SOU is so important to a whole-of-organization approach to learning and teaching. In Chap. 1 , we described the problematic nature of the current architecture and social construction (Conway 1968) of the pre-contextual university based on its inability to exert whole-of-organization influence on how teachers teach and students learn. In Chap. 2, we built a new set of commitments and a policy framework for learning and teaching. For many leaders used to a pre-contextual change approach, the next step would be to invoke the effort chain to deliver the commitments and policy. This means first calling in the professional developers, then redesigning awards and grants, setting up secondments, new roles, and changing the organizational chart. In the preceding chapters,

A. Bain and L. Zundans-Fraser, *The Self-organizing University*,
DOI 10.1007/978-981-10-4917-0_3

we explained and shared examples showing why such an approach does not work to build whole-of-university learning and teaching capacity. This chapter describes a different course of action.

3.2 Scope, Depth, Derailers, and Co-evolution: The Case for Embedded Design

According to Dede (2006, 2016), an innovation needs to possess the requisite scope and depth required to be scalable. Most change processes in education are dependent on altering the way educators engage in their normal work—how they think act and feel about what they do every day in their professional lives. While the specific target of an innovation may be a new approach to assessment, curriculum, teaching, or technology, the support required for successful implementation at scale is frequently of much broader scope and depth than that usually assumed by change agents. All too often, change efforts target the immediate conditions required to make the change happen (e.g., providing professional development to upskill faculty members in an assessment, curriculum or teaching approach, or ensuring a new technology is in place). This approach fails to recognize the way change impacts individual faculty members more broadly and the scope and depth of effort required to support a change at scale.

In Chap. 1, we saw a pretense of scope in the elements associated with the pre-contextual university's effort chain (e.g., the awards, grants, standards, promotion frameworks, etc.) that addressed a range of learning and teaching-related activity. However, the superficial treatment of those elements, their lack of connectedness, and basic incompatibility with whole-of-organizational change rendered them unable to exert whole-of-organization influence. For example, promotion frameworks lacked the specification to engender the kind of workable distinctions necessary to make promotion decisions valid and reliable. Standards, when viewed closely were too general to drive determinations of quality and productivity. Awards did not possess trajectory in terms of exerting a generalized influence on the quality of teaching across the university.

We also witnessed little interest in building a deep understanding of the learning and teaching context in the pre-contextual university. Scalable innovation in learning and teaching requires a deep understanding of professionally controlled practice. However, taking this first step to understand what learning and teaching means is almost never on the radar in the effort chain approach (Bain and Drengenberg 2016) . For example, we found little evidence in a national study of the leadership of learning and teaching in higher education (e.g., Scott et al. 2008) to indicate leadership roles and judgments of leadership competence at the highest level in universities were driven or informed by the existence of comparable and visible professionally controlled practice at scale. In fact, an empirical study by

(Scott and Scott 2015) suggest that many academics are either unaware of or unconvinced about whether such practice exist or even needs to:

> In describing what was effective teaching the prevailing view was that good teaching was directly related to having a doctoral qualification in their research discipline…. *"all you need to be a good teacher is to have your doctorate."* (Scott and Scott 2015, p. 8)

This quote indicates that building capacity with professionally controlled practice is not viewed as a need or goal for many academics. It is also a cautionary reminder of the distance higher education needs to travel to professionalize learning and teaching at scale. William Massy an experienced economic reformer of higher education offers the following insights related to this issue:

1. It's never wise to expect faculty to adopt a reform that appears alien to what they believe and know.
2. Faculty and academic leaders don't know as much as they think they know about how improvements, even ones that are mutually desired, can be achieved. (Massy 2016, p. 5)

If a university wants to change the way learning and teaching happens it must first recognize and consider the high likelihood of low commitment to, and perceived need for, a professional whole-of-organization approach. Second, any effort to respond to this low baseline must possess the meaningful scope and depth required to move a whole community in a new direction. That scope and depth may extend beyond what the community knows and feels about organizational improvement. Such support extends well beyond skill-building, to the systems, methods, tools, that influence the many ways a community will think, act, and feel about learning and teaching in a time of change. Our point here relates to the symmetry that needs to exist between what will be asked or required of a community and what will be offered in support of that request or requirement. The writing on educational change is replete with strategic and tactical advice including admonitions about resistance to change, the role of leadership, communication, collaboration, change models—their phases and stages. Although, to get to the starting line with whole-of-organization change, any process irrespective of its strategic and tactical trajectory must possess the scope and depth to achieve its goals.

Massy (2016) identifies the following as derailers of improvement in higher education. Most obviously, they serve as a useful benchmark for identifying what can go wrong in a change process. More importantly, they also provide useful insight to the scope and depth required of a change if the derailers are to be avoided:

1. Conventional mental models that cause the rejection of new ideas and block adoption processes;
2. Lack of closure that causes endless debate without a plan of action;
3. Lack of engagement that causes implementation to falter;
4. Lack of feedback that prevents participants from receiving maximal benefit;

5. Lack of follow-through on initial efforts or pilot projects and in commitments against performance;
6. Lack of recognition of those involved to celebrate successes. (Massy 2016, pp. 65–66)

The corollary of this list would indicate that a serious change process must be able to alter conventional mental models, engender widespread engagement, possess the mechanisms to generate timely feedback, create extant processes for reward and recognition, and include the kind of purposeful approach to implementation that generates widespread engagement and follow-through. If a change is to be respectful of and responsible for its impact on a community, it needs to constitute a responsible course of action. To be responsible, a whole-of-organization approach to learning and teaching needs to possess the scope and depth capable of addressing the aforementioned derailers as well as delivering its intended outcomes. As we saw in the preceding chapters, this is never the case in the pre-contextual university. In fact, the social construction and architecture of the pre-contextual university can be seen as a symmetrical accommodation of the derailers of change identified by Massy (2016), where an effort chain approach engenders an appearance of innovation without disrupting the equilibrium of its pre-contextual conditions and invoking the derailers. The latter would occur if innovation extended beyond the elective, selective, and exemplary approaches that are characteristic of the effort chain. Further, we contend that most efforts to change learning and teaching at scale in higher education would falter, even if they had "clear air and blue skies" (in terms of the people and process challenges of change implied by the list of derailers), because they do not possess the basic scope or depth required to deliver on their intended outcomes.

Transforming a pre-contextual effort chain approach to produce better outcomes for students involves paying serious attention to the derailers, along with the elements that comprise the way a university is constructed for learning and teaching, and the way that construction influences the totality of the professional life of its community. Embedded design is the approach for addressing scope and depth in the SOU to create the kind of mature effort that reshapes mental models for learning and teaching by embedding in the broader culture an approach that engenders the capacity for an organization to learn at scale (Massy 2016).

The idea of change as a process that includes many interactive elements is also a key to understanding the scope and depth of an innovation and the need for embedded design. According to Kauffman (1995), the emergence of successful innovation and change is a function of a series of dynamic coevolving conditions. Kauffman notes that the advent of the motor car as a scalable alternative form of transport did not simply occur because someone developed a new "buggy" with a workable internal combustion engine to replace a horse. The car as a form of mass transport came about because of new methods to produce better horseless buggies (cars) at scale and at an accessible price point. Widespread use also required an expanded network of roads with surfaces capable of supporting many cars. Those roads required the gas stations for replenishing the fuel source for those cars.

Mechanics were needed to fix broken cars, and even motels were required where people could stay when embarking on long trips (Kauffman 1995).

The popularization of the car as an alternative to horse and buggy required the co-evolution in depth and scope of a broad set of requirements for a new form of mass transport. A more recent example of this phenomenon relates to electric vehicles that require new and timely forms of refuelling if those vehicles are to be viable for long-distance travel and accepted as a scalable alternative to the internal combustion engine. The point here is that major social and technological[1] shifts require much more work than simply building out the core or primary innovation. They call for a network of broad and deep coevolving and reinforcing innovations and circumstances if they are to adapt dynamically to ongoing change.

We also know the efforts to border cross from old to new technologies (e.g., Weston and Brooks 2008) are made immensely challenging because these dynamic coevolving circumstances (the roads, gas stations, mechanics, etc.) rarely emerge as part of a smooth and timely transition. They all too frequently include many false starts, failures, gaps, and incomplete solutions before a new technology emerges at scale as a robust alternative to the prevailing approach. Further, in the early stages of an innovation the development of the many coevolving elements fall to the developers or change agents in the absence of the kind of dispersed control and modularization that occurs when an innovation is more mature (Horn 2014).

The difficulties experienced in border crossing to new technologies gives resonance to claims made by advocates of the status quo and critics of new approaches who find in the frequently cumbersome and fraught nature of change, many opportunities to amplify the incomplete nature of new technologies and highlight problems and inadequacies that serve to reinforce their status quo position. Many of the derailers identified earlier exist as a consequence of the coevolution problem in major change efforts. Coevolution signifies the dynamic and frequently unpredictable circumstances of change and innovation. It also reinforces the interactive nature of the many elements in a major innovation and reflects the challenge of working with all those moving parts. Coevolution suggest that not only does an innovation need to possess scope and depth but the things that make up the innovation need to be recognized as a dynamic complex system.

The SOU approach recognizes that like the advent of the motor vehicle, profound change in learning and teaching is not simply about taking up a new approach to assessment, curriculum, teaching, or technology. Each and all of those innovations require coevolving embedded design across all of the key elements required to support change for every member of the community. It is the thoughtful consideration of what those elements are and the ways they interact and coevolve in support of individual faculty members and the organization as a whole that drive whole-of-organization impact. In the SOU, commitments, policy technology, feedback, governance, etc., are the elements in that complex system. The examples

---

[1]We use the term technological here to denote innovation in the broadest sense and not restricted to information and communication technology.

included in this chapter will show how such a system can in part be an act of design including the ways technology can drive policy; how student feedback can instantiate better practice, and how the promise of workable distinctions can lead to the development of a valid and reliable framework for promotion. Embedded design in the SOU can be seen as an effort to take up the purposeful development of those elements and the way they interact and coevolve.

3.3 Defining Embedded Design

Embedded design means embedding an organization's commitments and policy in all of its systems, methods, and tools for learning and teaching and all in each. The goal: to create a learning and teaching context at scale. Embedded design replaces the effort chain architecture and social construction of the pre-contextual university with a comprehensive framework for self-organization. This includes connecting commitments and policy to the role of technology, governance and role, recognition, and reward. The approach does the following:

1. Instantiates[2] the learning and teaching context in all aspects of an organization's design, i.e., its technologies, its systems of reward and recognition, approach to professional development and governance;
2. Makes possible attributable relationships between normal practice and student outcomes at scale;
3. Builds self-reinforcing demonstrable connections across the elements (e.g., technologies, promotion frameworks, etc.) in routine normal work to produce a whole-of-organization schema for learning and teaching;
4. Creates the conditions for the emergent feedback required to establish standards and workable distinctions in the quality of learning and teaching.
5. Is capable of addressing the derailers of improvement efforts in higher education.

Figure 3.1 represents the way in which the learning and teaching context described in Chap. 1 and represented here at the center of the figure is instantiated by the design elements associated with the embedded design process. Embedded design focuses on paying attention to the many moving and interacting parts that impact the professional lives of faculty members and the university as a whole.

Our enthusiasm for embedded design is based upon longitudinal study of the application of the principle to program and organizational design in both compulsory and higher education. We have studied the embedded design principle in a range of contexts showing how it can improve professional knowledge, schema development; use of professional language, collaborative process, self-efficacy,

[2]Instantiation means, creating a real instance of an abstraction by producing a more detailed and refined version of a process, object or event (Dictionary.com, 2016).

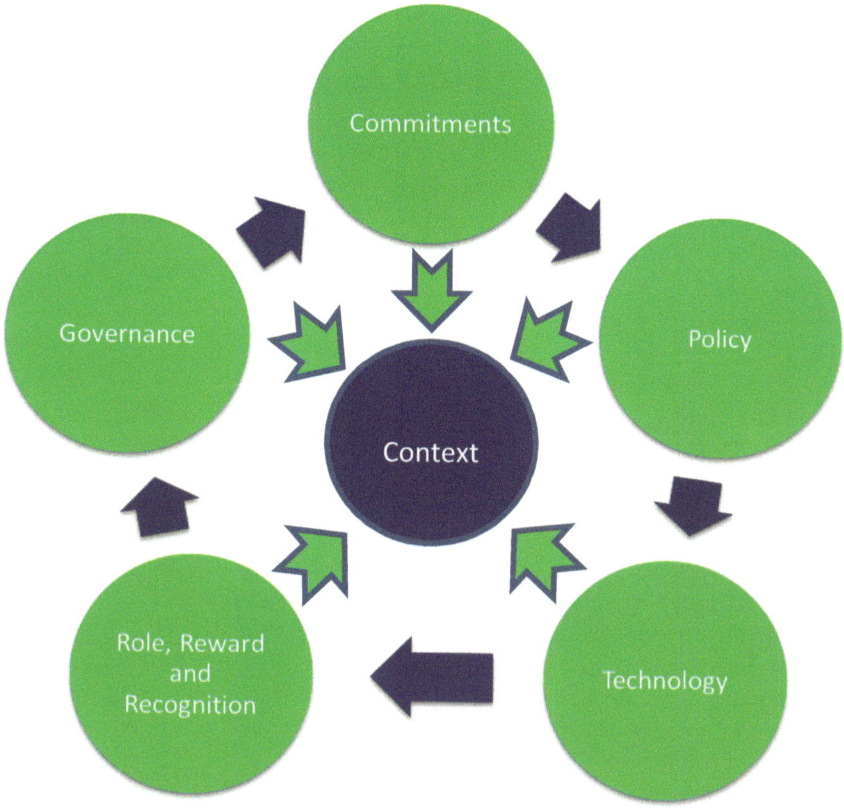

Fig. 3.1 Embedded design instantiating the learning and teaching context

enhance the translation of theory to practice, and the quality of teaching practice (Auhl, in progress; Bain 2007, 2010; Bain et al. 2009; Bain and Zundans-Fraser 2016; Bain et al. 2009; Grima-Farrell 2012; Lancaster 2016; Lancaster and Bain 2010; Zundans-Fraser and Lancaster 2012). The day-to-day engagement with the elements of embedded design described in Fig. 3.1 not only instantiate the learning and teaching context, it sets in place the conditions for the development of a shared schema for learning and teaching. When a faculty member engages with commitments and policy every time they use design, tools, share feedback, engage with the university's governance process, or consider promotion, they build a shared schema for what learning and teaching means. The examples that follow will bring additional clarity to the conceptual representation depicted in Fig. 3.1.

3.4 Embedded Design in Action

In Chap. 1, we presented a conceptual model describing the elements of the learning and teaching context in a self-organizing university. As previously stated, the goal of embedded design is to make real in practice the conceptual description of the learning and teaching context. This means creating the coevolving systems, methods, and tools necessary to produce attributable relationships with student outcomes, workable distinctions in performance, demonstrable standards, and emergent feedback.

In Chap. 2, we began the embedded design process demonstrating the way policy instantiated learning and teaching commitments at the SOU. The general and more abstract nature of the high level commitment was defined in more detail as part of policy. In the examples that follow, we show how the intentions of commitments and policy can be instantiated by embedding the tenets of policy in the design of software used in day-to-day normal work for developing learning experiences; how embedded design can drive the coevolution of the way students share feedback, and how commitments, policy, student feedback, and technology can drive the development of a valid and reliable promotion framework. A central feature of each example is the way the bottom-up methodologies of crowd-sourcing, commons-based peer production echo in the SOU approach. In addition, in Example 3.1 we introduce agent-based design as a further example of the bottom-up methodologies for building whole-of-organization capacity. In the example we show how agent-based design can be employed to design software for learning and teaching.

3.5 Agent-Based Software Design

Agent-based design involves describing a system from the perspective of its constituent units in order to understand the interaction of key elements (Bonabeau 2002). The approach focuses on sampling and modeling the interaction of agents in a real working context to understand the topology of that work and the rules and relationships that apply, including the way those agents engage in networked collaboration with their peers (Macal and North 2010). This information can then be used to drive software design. Most models of agent-based design involve an analysis of the need or problem, the situation or context and then the development of tools or process required to solve the problem (Bain and Drengenberg 2016; Billari et al. 2006; Macal and North 2010; Doran 2006). In Example 3.1 we show how the policy described in Chap. 2, was embedded in software tools for the day-to-day normal work of learning and teaching using an Agent-Based Design approach at Grandbridge SOU.

Example 3.1 Embedding Policy in Learning Design Tools at Grandbridge SOU

The leadership team at Grandbridge SOU knows that one of the community's biggest challenges is translating commitments and policy into a day-to-day working reality at Grandbridge SOU. The team recognizes the potential of technology to make the design of evidence-based practices more accessible for the community especially where many faculty members have limited professional knowledge about learning and teaching. The leadership team is also keen to extend its participatory approach, (first deployed in commitment and policy development) to the design of software tools for enacting the Grandbridge SOU policy. The Information Technology Division (ITD) at Grandbridge SOU assumes responsibility for implementing a bottom-up process to reflect policy in practice.

The team begins its work with Grandbridge SOU's policy statement (Figs. 2.3 and 2.5) about cooperative learning (CL) and active lectures (AL). The statement includes guidance about the way students work together and their accountabilities in cooperative learning, and using worked examples and guided and independent student practice to create active lectures (Hattie 2008). These features of both CL and AL are known to produce attributable effects on student learning. The policy statement provides a basic topology for key features to be included in the design of a tool that would help faculty design cooperative learning experiences and active lectures.

The ITD team initiates an agent-based design approach to analyze the design situation in three ways. First, they review the policy statements and examples of the policy in practice. Second, they ask members of the faculty who already know about CL and AL to explain their process for developing their learning experiences. The faculty members are asked (individually) to self-talk or narrate the process they undertake to design CL and AL using an example. The ITD team members gathered the examples as a permanent product record of the design process. The self-talk along with an examination of the products, are used to analyze and interpret the process, and in order to expand and clarify the topology for the purposes of software design. The team looked for consistencies and differences in the responses to the design tasks. The experienced developers were also asked what kind of feedback they would like to receive about their designs and the target areas for that feedback.

The first phase situational analysis identified common themes and processes that emerged from the team members' design efforts. The ITD team found some variability in the way each faculty member interpreted CL and AL although they also found that for the most part team members employed the key research-based features of the approaches when developing their learning experiences. Those key features echoed in faculty member responses to the questions about feedback. They wanted to know whether their examples represented the key features effectively and whether content and pedagogy were effectively integrated in their work. The experienced faculty members also expressed interest in receiving feedback while they were developing their experiences as well as emergent feedback on the design of the experience overall.

The ITD team built a composite description of the process using the input derived from the initial situational analysis and then asked a group of less experienced faculty members to respond to a design scenario. This second phase of the situational analysis involved explaining to the less experienced faculty members why the approaches were important and the benefits to students by drawing attention to the key features AL and CL. The faculty members were then asked three questions about the description. The first question asked for an open-ended reaction to the process to tease out opportunities and barriers to engagement with evidence-based practice; a second question asked about any elements of the model process that respondents would like to use; then responses were elicited about things that seemed difficult or cumbersome and how the process articulated in the composite descriptions could be made easier to use.

After completing the multi-level situational analysis (i.e., experienced and inexperienced developers) the IDT was ready to engage a software development team to create a software design model for CL and AL approaches. The design focused upon four key drivers: First, the extent to which the key commitments and policy elements associated with CL and AE were represented. Second, the representation of the design task using interface metaphors that would support both experienced and inexperienced designers. Third, making feedback an important part of the process in ways that would reflect the priorities of the policy and the needs described by both experienced and inexperienced designers. Figures 3.2 and 3.3 describes the basic layout an AL[3] design tools that shows how the agent-based design process is reflected in the layouts for a tool to assist in the design of active lectures.

In this layout, the developer is building an AL for a course about program evaluation and specifically how to define the object of the evaluation. The fields on the layout under the heading of learning activity help the user to scaffold key research-based elements of the experience. The field in the middle builds the content to be covered.

Figure 3.3 describes a layout where the instructor takes each part of the evaluation object and builds an example for demonstration and for student practice. Figure 3.4 describes the way a colleague can leave emergent feedback while their peer is developing a learning experience. In this case, the developer is receiving feedback about the design of an example they have built in the software.

Figure 3.5 describes feedback that includes a quantitative rating (on the slider). In this case, the feedback and the clarity of guidance provided to students about an assessment task. The colleague leaving feedback records a comment and a rating.

Figure 3.6 shows how students can leave feedback using a smartphone. Note the way the feedback is referenced to the research-based characteristics of the AL; in this case, how well the topic was introduced and the effectiveness of the worked examples. The different forms of feedback reflect the requirements shared by experienced users in the agent-based design process.

We can unpack Example 3.1 using the key features of the embedded design process described at the beginning of the chapter. First, we can see the way the software makes the concepts and big ideas of Grandbridge SOU's learning and

---

[3]Examples here are for active lecture. The cooperative learning designer features in Example 7.2.

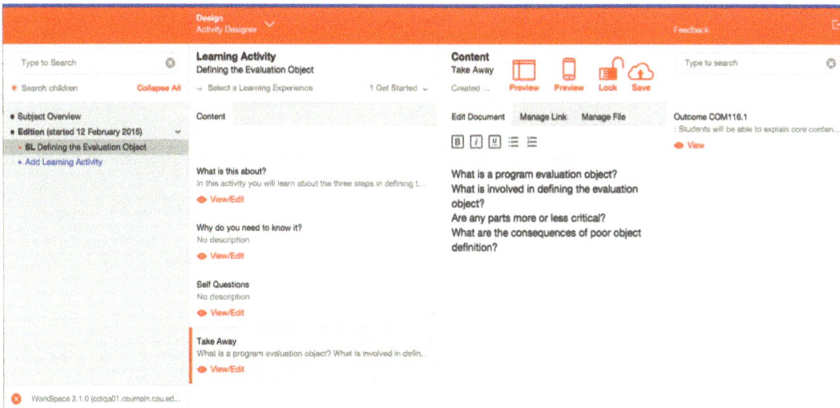

Fig. 3.2 Layout for designing an active lecture. (Republished with permission of Routledge, from Transforming the Measurement of Learning and Teaching in Higher Education, Bain and Drengenberg 2016; permission conveyed through Copyright Clearance Center, Inc.)

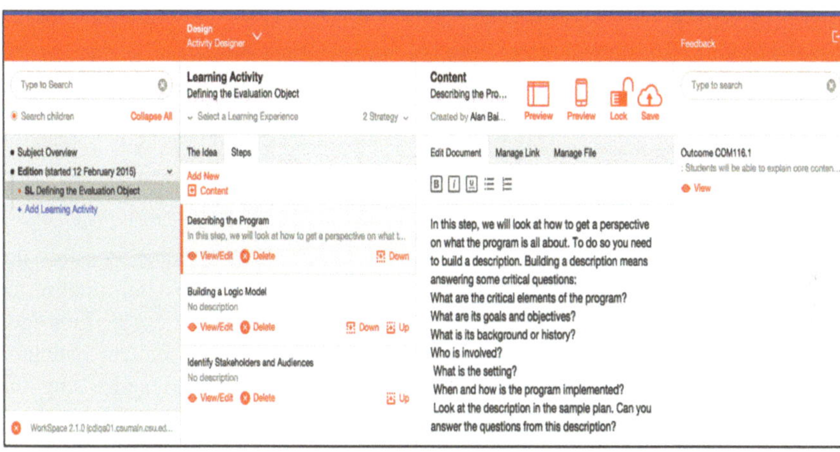

Fig. 3.3 Active lecture layout for building a worked example. (Republished with permission of Routledge, from Transforming the Measurement of Learning and Teaching in Higher Education, Bain and Drengenberg 2016; permission conveyed through Copyright Clearance Center, Inc.)

teaching context real in day-to-day normal work. The tools make the challenge of professionally controlled practice more accessible. Second, the software in its fields and interface requirements express the extensive track record of CL and AL to improve student achievement (Hattie 2008, 2015; Springer et al. 1999) and in doing so instantiates the attributable relationship between best practices and student outcomes. Third, the software develops the connections among the elements in the design. A commitment to evidence-based practice is defined in policy and instantiated in the design of software used for day-to-day work. Fourth, the feedback

Fig. 3.4 Emergent Peer feedback for active lecture. (Republished with permission of Routledge, from Transforming the Measurement of Learning and Teaching in Higher Education, Bain and Drengenberg 2016; permission conveyed through Copyright Clearance Center, Inc.)

generated by the software, described in Figs. 3.4, 3.5 and 3.6 makes possible the development of standards for the design of CL and AL and the determination of workable distinctions associated with the work completed in the tools. This idea will be developed further in Example 3.3 and illustrated in subsequent chapters. Fifth, the example can be seen as a significant step in terms of addressing the derailers of improvement efforts in universities.

In instantiating the learning context, the tools enable Grandbridge SOU to take a major step toward building new mental models (schema) for learning and teaching. They contribute to making practical a new schema for learning and teaching. Instead of creating arguments about fine points, the tools also provide a methodology for embracing the details of a change; in this case the transactions involved in evidence-based practice that contributes to student achievement.

The tools also create a pathway for engagement and the opportunities for sharing meaningful feedback and in doing so create the opportunity to recognize new efforts. The tools in their design and development and use also demonstrate follow-through from commitments to action and performance against commitments based upon the feedback they generate. Example 3.3 will show how feedback can extend to more formal recognition of effort through the development of a promotion framework. Finally, it is important to point out the way in which agent-based design

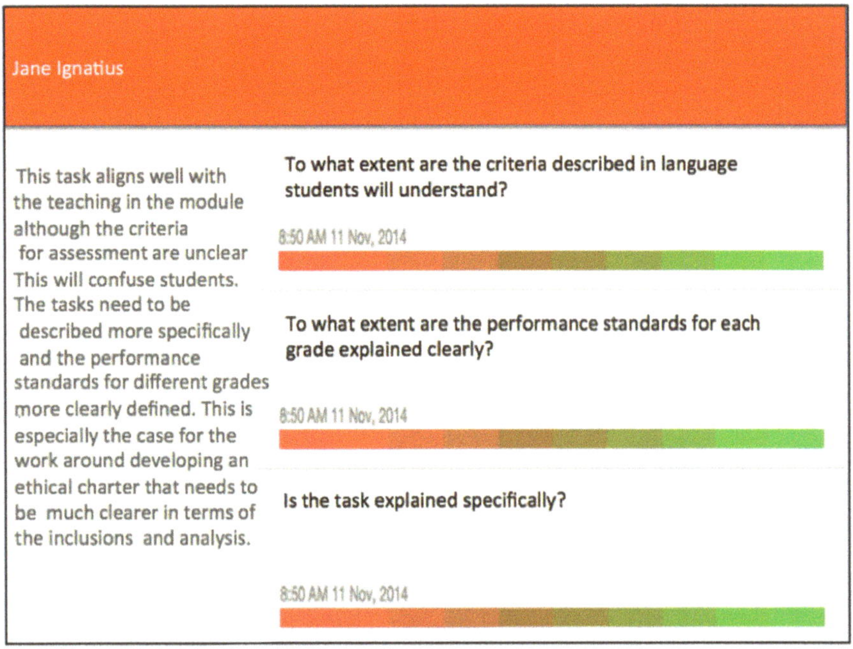

Fig. 3.5 Emergent Peer feedback for an assessment rubric. (Republished with permission of Routledge, from Transforming the Measurement of Learning and Teaching in Higher Education, Bain and Drengenberg 2016; permission conveyed through Copyright Clearance Center, Inc.)

shares a bottom-up and emergent quality with crowd-sourcing, and CBPP. The software design topology, situational analysis, and role of feedback are each developed by consulting members of the community to build the functionality of the tools bottom-up as a response to the needs and demands of those who will use them in their normal work.

The full scope of software development required to enable learning and teaching practice in the SOU approach is taken up in Chap. 7 as it relates to program design, program mapping, learning design assessment, and feedback. Example 3.1 shows in one area (designing cooperative learning experiences), the way evidence-based practice can be embedded in tools to produce workable distinctions in learning and teaching that can ultimately lead to the determination of the valid and reliable standards of practice. It is also important to note that we are not conflating the *existence* of the software with *using it* at scale to achieve better learning design. The existence of commitments, policy and tools represent a beginning in this regard although more needs to be done to create the coevolved and self-reinforcing circumstances necessary for professionally controlled practice at scale.

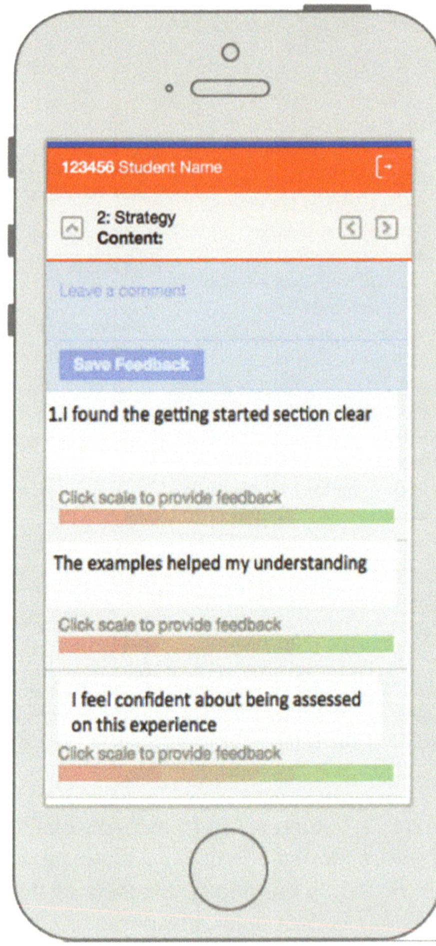

Fig. 3.6 Student feedback for a smart lecture. (Republished with permission of Routledge, from Transforming the Measurement of Learning and Teaching in Higher Education, Bain and Drengenberg 2016; permission conveyed through Copyright Clearance Center, Inc.)

3.6 Student Surveys Revisited

In Chap. 1, we were critical of student evaluation surveys when used in the pre-contextual university. We indicated that without an understanding of the learning and teaching context, those surveys can be little more than an expression of satisfaction with an idiosyncratic and autonomously constructed notion of learning and teaching practice. We also noted the preponderance of questions about instructor disposition (e.g., whether an instructor was enthusiastic) over questions about a professional practice (e.g., whether lectures were active and included practice opportunities). Example 3.2 revisits the survey issue showing how embedded design can be extended to the student survey because of the work undertaken at Grandbridge SOU. The example also shows how feedback from students needs to coevolve with the university's new program and the way that evolution does not always occur in a timely manner.

Example 3.2 Building a Different Kind of Student Evaluation Survey

John is an Associate Professor in the School of Veterinary Science at Grandbridge University and has been an active contributor to the Grandbridge SOU effort. He has recently completed the initial professional development modules required to participate in a pilot of the Grandbridge SOU approach. As part of the pilot, John has redesigned his Animal Genetics course in the new Grandbridge SOU learning and teaching software. He has just taught the course for the first time and has received his student evaluations.

John is frustrated. Overall, the feedback from students has been very positive. However, having just gone through an extensive engagement with a new approach to learning and teaching he feels the feedback he received does not reflect the immense time and effort he spent improving the quality of his teaching.

A faculty member known for being personable and setting high standards for his teaching, John's teaching has always been well received by students. However, his active involvement in the development of Grandbridge SOU has caused him to view his teaching differently. John is developing a new understanding of professionally controlled teaching practice and the effect it is having on student outcomes. He is building a new mental model, a new schema for his teaching and is excited by what he is accomplishing. However, he sees a big disconnect between what students are doing and achieving in his classes and the questions described in the student evaluation survey of his teaching. John has always recognized and continues to value highly the importance of his relationship with students, his enthusiasm and responsiveness. However, he also recognizes that the profound change in his professional practice is not represented in the feedback provided by the survey. He feels that the current survey has not evolved in step with the changes at Grandbridge SOU. The items lack a focus on those things he is doing to drive better achievement in his classes. In response, he submits a series of items for consideration by the leadership team and broader community. John's list includes the following:

1. Lectures included worked examples of key ideas;
2. I had an opportunity to practice new skills taught in lectures;
3. Cooperative group activity included clear roles and responsibilities for each student;
4. Assessment tasks in cooperative group activity made me individually accountable;
5. The performance levels required for assessment tasks were clearly described;
6. The distinctions between levels of performance (e.g., A, B, C) described in assessment rubrics were clear.

John has noticed a positive difference in student performance as a result of his redesigned course. He is having a firsthand experience exerting an influence on the 25% of the variance in student achievement contributed by the quality of instruction (Bloom 1976). John wants the evaluation of the course to include those things he is doing that make a difference for his students. His interest in doing so extends well beyond the recognition of the effectiveness of his redesigned course or his own practice. He does not want to get rid of the items about his effort and disposition. However, he knows that if the course and his teaching are to improve further, he needs to ask students for information about those things that are driving quality and student outcomes. He also knows that if Grandbridge SOU students learn more about those things that contribute to the variance in their achievement, they will become more effective communicators about the things that matter and in doing so contribute in a much more meaningful and partnered way to the quality of learning and teaching at Grandbridge SOU. Learning and teaching at Grandbridge SOU will co-evolve with better student feedback. Such feedback will instantiate existing strengths in the Grandbridge SOU design and point out areas that need to change and improve based upon the perspectives shared by students about their learning and teaching.

Example 3.2 is illustrative of a number of big ideas about embedded design in the SOU approach. First, we see the way in which the Grandbridge SOU effort is driving John's bottom-up contribution to the evolution of the design. John has crossed a border to a new way of thinking about learning and teaching. He is developing a new schema for his professional practice. As a consequence, he has identified a disconnect between the Grandbridge SOU approach and the feedback being requested of students. The existing survey has not coevolved with other aspects of the Grandbridge SOU embedded design. The student feedback survey reflects the effort chain pre-contextual paradigm from which Grandbridge SOU is emerging. From John's perspective, much of the effort chain survey is instantiating a model of practice that from his perspective no longer exists at Grandbridge SOU and as such is an anomaly. Further, John sees the incongruity between his teaching and the survey as a lost opportunity to build student understanding of and engagement with what is happening at Grandbridge SOU. He believes thatcourse evaluations should ask students to reflect about the things that are most important for learning, teaching and student learning outcomes. Grandbridge SOUs commitments, policy, tools, etc., have established what those things mean. In its pre-contextual form, the survey represents a model of learning and teaching that fails to fully reflect and value the ways Grandbridge SOU is changing. John's feedback is making possible the kind of bottom-up contribution to the design of the organization intended by the embedded design process. His new items will be shared and processed using the bottom-up methodologies described in this chapter and Chap. 2 consistent with the way self-organizing agency is expressed within the Grandbridge SOU community.

3.7 Valid and Reliable Promotion Frameworks

In the preceding chapters, we stated that the SOU pivots on the existence of professionally controlled practice at scale (Bowker and Star 2000). We saw in Chap. 1 and specifically in the description of the learning context how the existence of such practice makes attribution and workable distinction in learning and teaching possible. We also witnessed in Chap. 1 the illusory nature of promotion frameworks when they cannot be anchored in quality practice and genuine workable distinctions. With commitments, policy and tools student and other feedback in place, it becomes possible to generate the workable distinctions in learning and teaching practice necessary for a legitimate promotion framework. Example 3.3 describes the way commons-based peer production (explained in the last chapter) can be employed to extend the embedded design process to produce a promotion framework for Grandbridge SOU.

Example 3.3 Workable Distinctions and the Promotion Framework at Grandbridge SOU

We saw, in Example 3.2 the way John's deeper understanding of professionally controlled practice, made him aware of the limitations associated with the effort chain approach for recognizing the quality of learning and teaching. John identified a mis-match between his professional practice and the legacy pre-contextual student survey at Grandbridge SOU. The leadership team recognizes that John's frustration will soon extend to promotion and become a derailer if Grandbridge SOU retains the kind of pre-contextual promotion framework described in Chap. 1. The team must ensure that the community's commitment to workable distinctions in professional practice is reflected in the way Grandbridge SOU approaches reward and recognition. The ways faculty members are supported and recognized in learning and teaching need to be embedded in the Grandbridge SOU design and co-evolve with the changing nature of normal work and the effort expended by faculty members to build new professional knowledge and skill. As Grandbridge SOU progresses with its embedded design process, the conditions are emerging for the design of a new framework for promotion based upon the Grandbridge SOU commitments, policy and tools, and the availability of the feedback required to produce the workable distinctions that underpin a legitimate career progression in learning and teaching.

The leadership team employ the crowdsourcing steps of orientation, ideation and evaluation first described in Chap. 2 Aitamurto and Landemore (2015) to identify key areas and criteria for promotion in learning and teaching. To orient the community, the team asks all Deans and Department Heads to conduct workshops with faculty on the SOU work completed so far in the embedded design process. This includes a review of both the process and product outcomes of the embedded design to date including commitments, policy and the software tools. The product of the work to date is also made available through the Grandbridge SOU website, described in Chap. 2. The leadership team also requests the Human Resources Division post on the website examples showing the structure of promotion frameworks from other universities.

In the second ideation phase, the leadership team asks stakeholder groups and individuals to employ the key developments in embedded design completed so far (commitments, policy, and tools) along with the website information to develop a list of promotion categories (and their definition) for learning and teaching. The crowdsourcing software[4] first used to identify commitments is employed to collate, map and deliberate on the criteria submitted by the different groups and individuals. The evaluation phase is also repeated, where the list of areas or categories is returned to the community for evaluative voting process where participants are asked to vote their top 4 areas. The responses are further analyzed and evaluated the HR staff to ensure that the categories are inclusive of all areas described in the commitments and policy. Four categories are developed. They are:

1. Learning and Teaching Design: All work related to the design of programs and courses using evidence-based practice;
2. Teaching Practice: All work related to the delivery of learning and teaching in all modes including the delivery of professional development within the university and elsewhere;
3. Leading, and Collaborating in Learning and Teaching: All work related to activity in providing feedback, working in and leading collaborative learning and teaching teams;
4. Innovating in Learning and Teaching: All work related to problem-solving learning and teaching issues and needs primarily at course, program, faculty and university levels and elsewhere.

[4]There are a number of commercially available collaborative software platforms that can be employed for compiling, sharing and analyzing crowdsourced data.

After developing the categories, the leadership team initiates a repeat of the CBPP process described in Chap. 2 to instantiate what the four categories mean in practice and the performance criteria associated with the categories. Figure 3.7 describes an example of performance from the learning and teaching design category that reflects what the community determined as a criterion for the policy statement related to evidence-based practice.

Example 3.3 echoes the five characteristics of embedded design and in addition shows the "all in each" aspect of the definition. First, the framework further instantiates the learning and teaching model by further clarifying what each element of the conceptual model means. Second, the criteria in the framework focus on those aspects of practice that exert an attributable influence on student outcomes. A demonstrable relationship between design feedback and student outcomes is stated explicitly in the criteria for performance at the Assistant and Associate Professor levels. They are the aspects of practice that are the focus of policy and are embedded in the tools that faculty members are using to build learning experiences. The criteria target the use of those practices. Third, the framework not only makes connections among the elements in the Grandbridge SOU design, we see the "all in each" dimension of the embedded design definition. All of the design elements discussed so far are expressed in the framework (commitments, policy technology). Promotion is made actionable by the feedback available in the technology tools that reflects the policy and the university's learning and teaching commitments. Fourth the framework is based upon workable distinctions in practice. The criteria for the quality of design described in Fig. 3.7 are made valid and reliable by the data that link design work to student outcomes. The framework also shows the interaction among the elements in the developing Grandbridge SOU design. The feedback functionality of the design tools

Level	Learning and Teaching Design Criteria *-Evidence-based practice*
Assistant Professor	Has designed a minimum of 20 learning experiences in the SOU design space and at least one course; has received feedback of 80%[12] on those designs[13] Provides informed feedback to peers reflecting the evidence-based features of the designs used and the effective integration of content and pedagogy. Receives feedback from students at the 80% level for all items pertaining to learning design on the Grandbridge SOU Student Evaluation Scale[14]. Can link the feedback to student performance.
Associate Professor	Meets all criteria for Assistant Professor. Has designed a minimum of 3 courses in the SOU design software. Can demonstrate a record of advanced feedback to peers at individual course and program level that shows evidence[15] of improvement in the quality of courses design in student feedback and performance over time as a result of the feedback.

Fig. 3.7 Example of promotion criteria for the quality of learning and teaching design

described in Figs. 3.4, 3.5 and 3.6 make possible the determination of the quality of a given learning design which can be triangulated with student feedback and performance as indicated in the criteria described in Fig. 3.7. The triangulation of faculty member, peer, and student feedback with student performance links design to student outcomes. The promotion criteria reflect standards for the use of evidence-based teaching that emerge from sharing feedback about professional practice. The criteria reflect workable distinctions in the quality of learning and teaching. Fifth, the example shows the way in which the evolving embedded design process addresses the derailers of improvement. The instantiation of commitments and policy in tools, student feedback and promotion instruments serves to continue to build out an alternative mental model or schema. It shows that the finer points of the innovation are echoed in the way faculty are recognized and rewarded. That recognition is based on constant feedback, while the representation of the innovation in promotion shows the commitment and follow-through from big ideas to a fundamental operational cornerstone of the professional life of the institution.

While valid and reliable triangulated feedback can legitimize a career progression system, its greatest power lies in its contribution to the ground level normal work from which it emerges as students and faculty members work together. The feedback described in Figs. 3.4, 3.5 and 3.6 is shared and triangulated all of the time for immediate use in improving the design of learning experiences. While the data generated by these feedback loops can roll-up to inform summative decision-making for promotion, determining course and program quality, and meeting accreditation requirements, the uses of feedback in the SOU are always an emergent expression of the input necessary to engage in the day-to-day normal work of designing programs, courses and learning experiences. As such, the product of the embedded design process is to instantiate workable distinctions in quality that inform what to do next (as opposed to working out what happened, which is the focus of so much evaluation) in day-to-day learning and teaching (Bain 2007). Shared emergent feedback is also indicative of learning and teaching as a collaborative activity. The repeated sharing of emergent feedback builds a shared understanding of the learning context and enables the schema development necessary for effective professional collaboration in a learning community. Such a community can offer both the professional and social support for its members when an organization is in change.

3.8 Professional Capacity Building Last but not Least

Professional capacity building constitutes another important target of embedded design with in the roles, reward, and recognition dimension. We have intentionally left a discussion of professional development until the latter part of the chapter because it is the last aspect developed in an SOU embedded design process. In the SOU, professional development expresses the full set of needs and demands that will be placed upon the learning community. As such, the design of professional development or capacity building efforts need to reflect the full scope of the change as expressed in the embedded design process. This is the opposite to most change initiatives where the professional development lever is usually pulled as a first step in efforts to innovate. In chapters one and two we witnessed the way the pre-contextual effort chain university uses professional development as a first strike panacea for its learning and teaching ills. The Pre-contextual Grandbridge learning and teaching plan was underpinned by large doses of professional development to drive improvement in its status as a provider of online learning. The problem with that well-intended initiative is Pre-contextual Grandbridge did not know enough about what learning and teaching means to have its professional development make a targeted whole-of-organization contribution to quality. We see broader evidence of this problem in the literature on professional development in education. A study by Cole (2012) found over 34,000 references in the literature to professional development, only 97 of which were focused on student outcomes. Cole indicated the literature shows overwhelmingly that most professional development is ineffective in bringing about improvements in teaching and student outcomes. While the focus of Cole's work was the compulsory sector, there is little to suggest that the expenditures in higher education are resulting in a different outcomes given the pre-contextual circumstances that prevail across the sectors. We also noted in Chap. 1 that the uptake of professional development opportunities in learning and teaching is frequently low (Probert 2015) and faculty members are often skeptical about the value of those opportunities. This is unsurprising given the research on the efficacy of professional development.

PD fails in the pre-contextual university because it is expected to be the silver bullet driver of change instead of filling a meaningful role within an embedded design. By this we mean professional development needs to be a role player embedded within a broader interconnected and supportive learning and teaching context instantiated by commitments, policy, technology, role, governance and career trajectory that function interdependently and coevolve to create a new schema for learning and teaching practice. Most often PD is expected to carry forward a loosely articulated pre-contextual design approach where there is insufficient attention paid to other key components; and when attention is paid, it is usually in a manner that fails to recognize the interdependencies required to support new ways of learning and teaching. The role of PD like the effort chain it serves is not clearly connected to the professional lives and expectations of faculty members and fails the relevance test for many faculty members. Under these circumstances, it is no surprise that PD has such an unimpressive track record in changing practice at scale.

In the SOU, professional development is purposively focused on the knowledge and skills required to engage in quality comparable and visible professional practice. Before John could redesign his animal genetics course described in Example 3.3, he received extensive professional development in the Grandbridge SOU approach. He learned how to design active lectures and evidence-based cooperative learning among many other things associated with the design. However, his professional development experience was not expected to be singularly transformational. It was a part of Grandbridge SOU's embedded design situated within a broader context of new commitments, policy, technology, role descriptions and career progression, and governance structure, each of which were sourced from the community and designed to function as a coevolving supportive framework that redefined the learning and teaching context and instantiated a new schema. Each member of the community brings their personal agency to an environment capable of professionally controlled comparable and visible practice at scale.

3.9 The Loose Coupling Myth

The embedded design process described in this chapter signifies greater connectedness among the key organizational elements that drive learning and teaching at scale in a university. The scope, depth, and coevolutionary nature of the embedded design process stands in contrast to the loose coupling (Weick 1976) that is the hallmark of the pre-contextual effort chain organizational structure of universities. Loose coupling is characterized by an absence of regulation, loose organizational arrangements, lack of consensus, and low levels of coordination (Weick 1976; Orton and Weick 1990). Since Weick's seminal work, many pre-contextual universities have worn loose coupling like a badge of honor invoking it as an effective organizational form consistent with effective adaptation to change, responsive decision-making, multiple decision-making pathways, a cultivator of institutional ingenuity, initiative, and flexibility (Orton and Weick 1990). These potentials may exist for organizations in fields that possess fundamental professional control. However, in an echo of Conway's Law, (Conway 1968) the extensive evidence about the inability of universities to determine what learning and teaching mean, the failure to establish what constitutes quality and productivity in relation to student outcomes, along with the inability to exert scalable influence on their core activity (learning and teaching practice) would suggest a symmetry between loose coupling in universities and its outcomes in the pre-contextual university. The touted benefits may reside more in the realm of myth than reality when applied to learning and teaching in higher education.

While many in the higher education sector interpret loose coupling as the optimal model for the dominant values of autonomy and academic freedom in higher education, in normal practice it simply serves to perpetuate the effort chain approach to learning and teaching especially when there is no underpinning comparable and visible professional practice to serve as a foundation for the touted benefits of the approach. The loose-tight coupling discourse also invokes a conceptualization of tight coupling as synonymous with command and control, top down inflexible organizational approaches, strict performance management criteria, management by objectives, and hierarchical management structures, (Weick 1982). In fact, this is the kind of management-oriented and more tightly coupled model that critics would suggest has pervaded higher education over the last two decades. This characterization stands in contrast to the way embedded design is articulated in the SOU, and as we will show in Chap. 5, from a governance perspective, is more consistent with the effort chain approach.

The connections made through embedded design function in a totally different role for a profoundly different purpose in the SOU—to create a visible and comparable understanding of what learning and teaching mean in order to disperse control and share feedback. For example, it is the shared understanding of evidence-based practice that makes the technology supported feedback described in Example 3.1 possible. That feedback is employed to generate workable distinctions in the design of learning experiences based on the bottom-up views and perspectives of peers engaged in the normal work of course design not the top down edicts of a command and control leadership model. The shift in John's understanding of his professional practice engendered by embedded design drove a bottom-up reconsideration of the items in Grandbridge SOU student evaluation survey. Embedded design creates the clarity of purpose and process, the shared schema, and shared understanding for self-organizing action as expressed in John's recommendations for new survey items. Embedded design is about building a deeper understanding of what is managed in an organization.

The tighter coupling in the SOU creates the support and functional cohesion (TheBojan 2015) required for an organization to function professionally at scale. In essence, it is the order required to replace top down control (Bain 2007) where there exists a shared understanding and schema for collaborative professionally controlled practice. These are the baseline conditions that exist in any complex adaptive system capable of generating the emergent feedback synonymous with the adaptation, flexibility and innovation assigned to those systems. Embedded design is about creating the kind of functional cohesion in an organization necessary to legitimately distribute leadership and engender genuine bottom-up self-organization. In effect embedded design creates the kind of high cohesion (TheBojan 2015) around professional control that makes the potential benefits of loose coupling possible.

So far in this chapter we have used examples related to technology, student feedback, and career progression to illustrate how embedded design creates the connections among elements in an organization's constitution for learning and teaching. Those connections serve to instantiate the learning context and drive in practical ways a new organizational schema for learning and teaching. In the SOU, the principle of embedded design illustrated in the examples is also applied to other areas including role descriptions, governance and the design of professional development and capacity building. When an organization understands what learning and teaching means in it, it can apply that understanding to all aspects of its design. The chapters that follow will elaborate on additional applications of the principle.

Takeaways

In this chapter, we described embedded design and its role in the SOU. The key takeaway propositions from this chapter are:

- A whole-of-organization impact on learning and teaching requires a whole-of-organization change process;
- Most change efforts in learning and teaching lack the scope and depth to exert impact at scale;

- A whole-of-organization change process must be capable of addressing the derailers of improvement (Massy 2016) in universities;
- Embedded design means embedding an organization's commitments and policy in all of its systems, methods, and tools for learning and teaching and all in each;
- Embedded design replaces the effort chain architecture and social construction of the pre-contextual university with a comprehensive framework for self-organization that translates a university's big ideas about learning and teaching into the systems, methods, and tools for routine professional practice at scale;
- Instantiating the learning and teaching context is the cornerstone of embedded design and the SOU approach;
- To be effective, professional development needs to be embedded in an instantiated learning and teaching context;
- The elements in the embedded design process coevolve to build a whole-of-organization schema for learning and teaching;
- Loose coupling expresses the effort chain organization of the pre-contextual university;
- Embedded design provides the order to disperse control and distribute leadership that makes self-organization possible.

References

Aitamurto, T., & Landemore, H. (2015). Five design principles for crowdsourced policymaking: Assessing the case of crowdsourced off-road traffic law in Finland. *Journal of Social Media for Organizations* 2(1), 1–19. Retrieved from http://www.mitre.org/public/jsmo/pdfs/02-01-5-designprinciples.pdf

Auhl, G. (in progress). The development of pre-service teachers' schema for teaching students with diverse learning needs.

Bain, A. (2010). A longitudinal study of the practice fidelity of a site-based school reform. *Australian Educational Researcher, 37*(1), 107–124.

Bain, A. (2007). *The self-organizing school. Next generation comprehensive school reforms.* Lanham, MD: Rowman & Littlefield.

Bain, A., & Drengenberg, N. (2016). *Transforming the measurement of learning and teaching in higher education.* New York: Routledge.

Bain, A., & Zundans-Fraser, L. (2016). *Rising to the challenge of transforming higher education: Designing universities for learning and teaching.* New York: Springer Verlag.

Bain, A., Lancaster, J., & Zundans, L. (2009a). Pattern language development in a pre-service learning community. *International Journal of Teaching and Learning in Higher Education, 20* (3), 336–349.

Bain, A., Lancaster, J., Zundans, L., & Parkes, R. J. (2009b). Embedding evidence-based practice in pre-service teacher preparation. *Teacher Education and Special Education, 32*(3), 215–225. doi:10.1177/0888406409339999.

Billari, F., Fent, T., Prskawetz, A., & Scheffran, J. (2006). *Agent-based computational modelling: Applications in demography, social, economic and environmental sciences.* Heidelberg: Physica-Verlag.

Bloom, B. S. (1976). *Human characteristics and school learning.* New York: McGraw-Hill.

Bonabeau, E. (2002). Agent-based modeling: Methods and techniques for simulating human systems. *PNAS, 99*(3), 7280–7287. Retrieved from http://todd.bendor.org/upload/PNAS-2002-Bonabeau-7280-7.pdf

Bowker, G., & Star, S. (2000). *Classification and its consequences*. Cambridge, Massachusetts: The MIT Press.

Cole, P. (2012). *Linking effective professional learning with effective teaching practice*. Retrieved from Australian Institute for Teaching and School leadership website: http://www.aitsl.edu.au/docs/default-source/default-document-library/linking_effective_professional_learning_with_effective_teaching_practice_-_cole

Conway, M. E. (1968). How do committees invent. *Datamation, 14*(5), 28–31.

Dede, C. (2016, February 9). Designing an educational innovation to achieve scale: Five critical concepts. *The Evolution: A destiny solutions illumination*. Retrieved from http://evolllution.com/managing-institution/operations_efficiency/designing-an-educational-innovation-to-achieve-scale-five-critical-concepts/

Dede, C. (2006). Scaling up: Evolving innovations beyond ideal settings to challenging contexts of practice. In R. K. Sawyer (Ed.), *Cambridge handbook of the learning sciences* (pp. 551–566). Cambridge, England: Cambridge University Press.

Dictionary.com. (2016). Definition of instantiation. Retrieved from http://www.dictionary.com/browse/instantiation

Doran, J. E. (2006). Agent design for agent based modelling. In F.C Billari, T. Fent, A. Prskawetz & J. Scheffran (Eds.), *Agent based computational modelling: Applications in demography, social, economic and environmental sciences*. Physica-Verlag (Springer), pp. 215–223.

Grima-Farrell, C. (2012). *Identifying factors that bridge the research-to-practice gap in inclusive education: An analysis of six case studies*. (Doctor of Philosophy), Australia University, Bathurst.

Hattie, J. (2015). The applicability of visible learning to higher education. *Scholarship in Teaching and Learning in Psychology, 1*(1), 79–81.

Hattie, J. (2008). *Visible learning: A synthesis of over 800 meta-analyses relating to achievement*. New York: Routledge.

Horn, M. (2014, July, 10). Unbundling and re-bundling in higher education. Retrieved from http://www.forbes.com/sites/michaelhorn/2014/07/10/unbundling-and-re-bundling-in-higher-education/#2b41d28879c2

Kauffman, S. (1995). *At home in the universe: The search for the laws of complexity and self-organization*. New York: Oxford University Press.

Lancaster, J. (2016). *The instantiation of theory into practice for preservice teachers in inclusive education*. (Unpublished doctoral dissertation). Australia University, Bathurst, NSW.

Lancaster, J., & Bain, A. (2010). The design of pre-service inclusive education courses and their effects on self-efficacy: A comparative study. *Asia-Pacific Journal of Teacher Education, 38* (2), 117–212.

Macal, C., & North, M. (2010). Tutorial on agent-based modelling and simulation. *Journal of Simulation 4*(3), 151–162.

Massy, W. (2016). *Re: engineering the university: How to be mission centered, market smart, and margin conscious*. Baltimore: Johns Hopkins University Press.

Orton, D. J., & Weick, K. E. (1990). Loosely coupled systems: A reconceptualization. *Academy of Management Review, 15*(2), 203–223.

Probert, B. (2015). *The quality of Australia's higher education system: How it might be defined, improved and assured*. Office for Learning and Teaching Discussion Paper. Australian Government Office for Learning and Teaching. Retrieved from http://www.olt.gov.au/resource-quality-australias-higher-education-system.

Scott, G., Coates, H., & Anderson, M. (2008). *Learning leaders in times of change: Academic leadership capabilities for Australian higher education*. Retrieved from http://research.acer.edu.au/cgi/viewcontent.cgi?article=1001&context=higher_education

Scott, D., & Scott, S. (2015). Leadership for quality university teaching: How bottom-up academic insights can inform top-down leadership. *Educational Management Administration & Leadership, 44*(3), 511–531. doi:10.1177/1741143214549970.

Springer, L., Stanne, M. E., & Donovan, S. S. (1999). Effects of small-group learning on undergraduates in science, mathematics, engineering and technology: A meta-analysis. *Review of Educational Research, 69*(1), 21–51. doi:10.3102/00346543069001021.

TheBojan. (2015, April, 8). High cohesion, Loose coupling. Retrieved from https://thebojan.ninja/2015/04/08/high-cohesion-loose-coupling/

Weick, K.(1976). Educational organizations as loosely coupled systems. *Administrative Science Quarterly, 21,* 1–19.

Weick, K. (1982). Administering education in loosely coupled schools. *Phi Delta Kappan* (June), 673–676.

Weston, M., & Brooks, D. (2008). Critical constructs as indicators of a shifting paradigm in education: A case study of four technology-rich schools. *Journal of Ethnographic and Qualitative Research in Education, 2*(4), 281–291.

Zundans-Fraser, L., & Lancaster, J. (2012). Enhancing the inclusive self-efficacy of preservice teachers through embedded course design. *Education Research International, 2012*. doi:10.1155/2012/581352

Chapter 4
Getting the Right Feedback for Learning and Teaching

4.1 Introduction and Overview

In this chapter, we explain new opportunities for gathering and using data in a self-organizing university (SOU). We define and describe an approach entitled emergent feedback (Bain 2007) and show how it is different from traditional notions of evaluation, performance monitoring, and acquittal on strategic plans.

To open the chapter, we show why pre-contextual effort-chain universities cannot get the data they need to improve learning and teaching at scale. We employ an example from Pre-contextual Grandbridge to illustrate the challenges associated with the effort-chain approach. We then build upon the work of preceding chapters to show how feedback emerges from the learning and teaching context in the SOU. Specifically, we describe how feedback becomes a deeply embedded part of routine normal work as opposed to something an organization does after-the-fact to find out what happened. Our second and third examples illustrate the emergent feedback approach in practice at scale within Grandbridge SOU. We conclude the chapter with a discussion of the implications of emergent feedback for higher education more generally including the ways emergent feedback makes implausible the divide between strategy and execution in leading universities; and the way such feedback can transform a university's learning and teaching productivity.

4.2 Emergent Feedback Defined

In the preceding chapters, we explained how understanding the learning and teaching context can transform the way a self-organizing university is designed. Making learning and teaching comparable, visible, and professionally controlled (Bowker and Star 2000) means that feedback can focus on things that are known to make a difference in student learning outcomes. As illustrated in Chap. 3, the

A. Bain and L. Zundans-Fraser, *The Self-organizing University*,
DOI 10.1007/978-981-10-4917-0_4

embedded design of governance processes, technology tools, roles and career progression means that feedback can be connected deeply to the normal work associated with learning and teaching in all facets of a university's operation. Emergent feedback can be defined as proximal data that emerges from the day-to-day professional work of learning and teaching in design, enactment and engagement (Bain 2007; Bain and Drengenberg 2016). The term proximal in the definition means close to the design and or delivery of professionally controlled practice in terms of context and timing. Emergent feedback has seven distinctive characteristics. Emergent feedback:

1. Focuses on and is a part of the key day-to-day transactions associated with normal work—visible and comparable professionally controlled practice (Bowker and Star 2000);
2. Emerges from and contributes to the agency of those involved in learning and teaching;
3. Is transparent and collaborative. It employs multiple perspectives about what is happening in the context;
4. Is enabled and empowered by technologies that articulate the learning and teaching context;
5. Is similar at scale (Mandelbrot 1982). A program team meeting to solve a course level issue related to the design of assessment tasks employs the same process and feedback as a university level team looking at faculty or university performance with assessment;
6. Exists to make a system or organization function in real-time, to inform immediate activity, problem-solve, and adapt over working out what happened;
7. Makes educational measurement proximal to learning and teaching practice and predictive of learning effects (Bain and Drengenberg 2016) .

4.3 Peer Moderation at Pre-contextual Grandbridge

In Example 4.1 we describe Pre-contextual Grandbridge's effort to build greater quality into its assessment process with peer moderation. We then employ the emergent feedback characteristics to unpack the example in order to understand the impact of the moderation process at pre-contextual Grandbridge.

Example 4.1 Jane's Peer Assessment Moderation

Jane is an assistant professor in the school of community health at Pre-contextual Grandbridge. She is reviewing a colleague's syllabus as part of the university's peer moderation process. Jane, like many of her colleagues has no formal background in teaching or assessment although she has attended professional development programs on assessment conducted by the Center for Excellence in Learning and Teaching (CELT) at Pre-contextual Grandbridge. Jane is struggling with the development of assessment items for her course and especially with the construction of assessment rubrics that adequately differentiate student achievement. She is reviewing the following questions on a moderation checklist resident on Pre-contextual Grandbridge's Learning Management System (LMS). She is required to complete the checklist for the assessment tasks described in her colleague's syllabus. The questions on the checklist include:

- Are the standards for assessment criteria clear? Yes/No
- Are the criteria for grades clearly differentiated? Yes/No
- Is the assessment task linked to course outcomes? Yes/No
- Is there an obvious connection with the learning and teaching in the course? Yes/No
- Is the task rigorous? Yes/No

Jane has read the syllabus carefully and wants to provide useful feedback although she is not confident about providing yes/no responses to the questions. She is being asked to provide definitive feedback about her colleague's work in areas that she feels are beyond her scope of expertise. She is conflicted, questioning her capacity with what she is being asked to do yet wanting to fulfil a responsibility to her colleague and the university. While the questions seem straightforward, Jane knows from attending the CELT workshop, the design and differentiation of assessment criteria require sophisticated knowledge and skill. While the workshop helped to build Jane's knowledge of assessment, it also served to clarify just how much she does not know about assessment design.

Jane is also unsure about the content standards expected of students in areas not necessarily within her scope of expertise. The feedback Jane has received about her assessment tasks has been mostly positive. Although she is uncertain about the legitimacy of some of the judgments made by her peers just as she is uncertain about the feedback she is providing. In instances where Jane has received corrective feedback about her assessment tasks, she is also unclear about how to make a number of the requested improvements. Jane's experience with moderation is echoing across Pre-contextual Grandbridge as all schools and faculties engage with the peer moderation process. The CELT is employing an analytics function of the new LMS to aggregate scores from completed checklists across the schools, faculties and the university overall. The aggregated scores will be used to report progress on the university strategic planning goals related to assessment.

What can we take away from Jane's efforts with peer moderation? First, Jane's experience is definitely about her day-to-day teaching work. However, that work has little to do with an understanding of visible and comparable professional practice and an understanding of the learning and teaching context from which such practice emerges. As we have seen in the preceding chapters, such an understanding does not exist at Pre-contextual Grandbridge. The capacity to participate in the moderation process rests on the CELT professional development program and there is a gap between its capacity building outcomes and the expectations of the moderation process. This makes it difficult for Jane when she is asked to judge the work of others as there are no standards, attributable relationships and workable distinctions on which to base her judgments. These are the same features that would have contextualized and empowered the CELT professional development. Jane will give feedback based on her idiosyncratic effort-chain experience and the learning derived from attending the professional development workshop. There is no possibility that her feedback will be indicative of known quality practice at scale or more specifically a shared understanding of assessment design.

Second, Jane is bringing the agency afforded by the effort chain and specifically the CELT professional development program to the moderation task. Jane's discomfort as a moderator reflects her belief that she lacks sufficient experience with the effective research-based learning and teaching practice required to answer the moderation question informatively. Jane's agency is an expression of Pre-contextual Grandbridge's architecture for learning and teaching. Jane is also unsure about the validity of the feedback she receives as well as what to do to act on the recommendations. We see in the example the way a well-intended peer mediated exercise in quality assurance can regress to the kind of connoisseurship first described in Chap. 1 in the absence of the professional knowledge required to engage with professionally controlled practice. The reality is that most members of the community are unsure about what to do.

Third, it is possible to view the peer mediated approach to moderation as collaborative and transparent. Jane is assisting a colleague who will get to see the feedback Jane provides. However, the feedback Jane will share is a product of the private-practice design work of her colleague that is held up for ex-post-facto summative judgment. There is little collaboration in the formulation of the assessment task or opportunity to provide the iterative feedback that could help shape her colleague's work when designing the task in the first instance. This includes connections between the task and the broader learning design for the course which is inextricably connected to the design of the assessment task in CBA.

Fourth, Jane is also using a technology to complete the task. She is completing the moderation scale using the survey functionality of Pre-contextual Grandbridge's new LMS. The CELT is using the analytics capability of the LMS to aggregate the scores of all surveys completed across the university. However, the use of technology is simply automating current practice. There is no sense that the LMS is providing a capability that helps Jane and her colleagues to better design and/or moderate assessment tasks; and not in a manner consistent with the examples described in Chap. 3 where the technology tools helped faculty members design better learning experiences.

Fifth, the university-wide aggregation of the moderation checklists by CELT is ensuring the data are scaling up. The leadership team at Pre-contextual Grandbridge will carefully examine the scores when reviewing the university level report produced by CELT. Conversation will focus on improving performance and acquittal against the strategic planning targets described in Example 2.1. The higher level leadership will assume a responsibility to act strategically about whatever the date suggests. However, the focus will be on the data not what it means. The report is not a scaled up representation of known quality assessment practice that is understood and self-similar across the community. It is as likely to represent an aggregation of the concerns and needs expressed in Jane's experience with the process. There is every likelihood that the members of the leadership team at Pre-contextual Grandbridge will share Jane's limited understanding of quality assessment. In fact, the quality that is most obviously self-similar at all levels of the Pre-contextual Grandbridge community is likely to be the limited understanding of evidence-based assessment practice.

Sixth, the leadership at Pre-contextual Grandbridge will no doubt seek timely action on the CELT report and specifically in relation to the timelines of the strategic plan. However, the report is not an expression of a methodology for using feedback all-of-the time to problem-solve ongoing needs and issues. There is no normal work process for building assessment tasks at scale from which such feedback could be derived. Whether seen as a formative waypoint in the strategic planning process or a summative indicator, any less than acceptable results will no doubt invite questioning about why the university's strategy for assessment is not being executed effectively. This will occur despite the fact that there is no articulated professionally controlled process for execution beyond the CELT professional development program. In the effort chain, none of the conditions are in place for implementing strategy at scale. The leadership will simply direct those responsible for execution to do better.

Seventh, Jane's engagement is an echo of the way Pre-contextual Grandbridge is constituted for learning and teaching. She provides feedback after-the fact. There is an expectation that Jane's colleague will employ the feedback although when and how this happens is not part of the process. The remedial work derived from Jane's feedback will occur at some time before the course is taught in its next iteration, although the only way for Jane to look at those changes is to request the moderation of the same course again next year. Further, the only way to check whether the moderation feedback is having an effect at scale is to analyze more moderation checklists or review student satisfaction surveys.[1]

Finally, the extent to which the feedback described in Example 4.1 is proximally and causally linked to action that will improve student performance is unknown at Pre-contextual Grandbridge. There is no shared understanding of the learning context, embedded design and schema for learning and teaching, and as a consequence the impact of peer moderation feedback on teaching practice and student outcomes is unknown.

4.4 Resolving the Very Very Problem with Emergent Feedback

In Example 4.2, we describe the SOU corollary to the moderation process applied to the design of assessment tasks and unpack the example using the same characteristics of emergent feedback.

Example 4.2 Jane Experiences the 'Very Very' Problem in Assessment at Grandbridge SOU

Grandbridge SOU is in the second year of implementing a new criterion-based approach to assessment (CBA) and the first year when all courses at the university are required to have a full explication of the grading criteria for every assessment task. Jane is now working in the Grandbridge SOU Coursespace tools developing the opening unit of her course on *Designing and Implementing Program Evaluations.* The unit is focused on explaining the object of a program evaluation. Jane is developing the grading rubric for the first

[1]The questionable validity of these measures was taken up in Chaps. 1 and 3.

assessment task and is having difficulty differentiating the performance criteria she is seeking. She is reviewing the grading rubrics for the assessment and needs some help (Fig. 4.1).

Jane invites her peers to take a look at her work and leave some feedback using the blogging capability of the course tools. She checks the feedback on the rubric for the assessment task and finds the following blog posting in the course tools (Fig. 4.2).

Jane's concerns about her rubric are confirmed by feedback from her colleagues. She recognizes the "very very" problem described in the first blog posting and sees that the rubric in its current form will be of little help to students. The second piece of feedback helps Jane to take some next steps. She does two things. She searches online for some published program evaluations and looks at their descriptions of the evaluation object. She begins to identify more and less effective descriptions. Second, she uses the examples to build out some much clearer and specific statements of what a description of the evaluation object should look like. Jane finds it much easier to describe different levels of performance from these more specific descriptions. She uses here new descriptions to revise the rubric and posts her work for more feedback.

We can use the seven characteristics of emergent feedback to unpack Jane's experience when developing her assessment rubric at Grandbridge SOU.

First, Jane is engaged in the day-to-day work of developing her course. She is asking for feedback from peers because she is having difficulty with "real time" course development as she builds an assessment rubric. The feedback is proximal or close to the need/problem Jane is facing in terms of context and timing. The feedback will assist her to improve the task and rubric immediately. It is this tight association between feedback and normal work that distinguishes the emergent feedback construct from notions of formative or summative evaluation (Scriven 1967) or evaluation more broadly. The latter refer primarily to making judgements when a program or process is either in development or at completion. Decisions are then made about change in design, continuation or termination, or success/failure usually against predetermined goals. The point here is that an emergent feedback approach is part of an embedded organizational design one that seamlessly integrates data, process, and technology to establish what to do next from a learning and teaching perspective (Bain and Drengenberg 2016). Emergent feedback is a part of an organization's operating structure, available all of the time. It is the fuel of a system's total operation driving its capacity to adapt and change all the time without serving a specific type of judgment or time-dependent decision (e.g., promotion, termination, success, or failure). While it is possible for such evaluative decisions to be abstracted from emergent feedback in reports for summative decision-making, they are not its reason for being.

The submission includes a satisfactory description of the four key elements of the evaluation object.	The submission includes a strong description of the four key elements of the evaluation object.	The submission includes an outstanding description of the four key elements of the evaluation object.	The submission includes an exemplary description of the four key elements of the evaluation object.

Fig. 4.1 Excerpt from assessment task Rubric

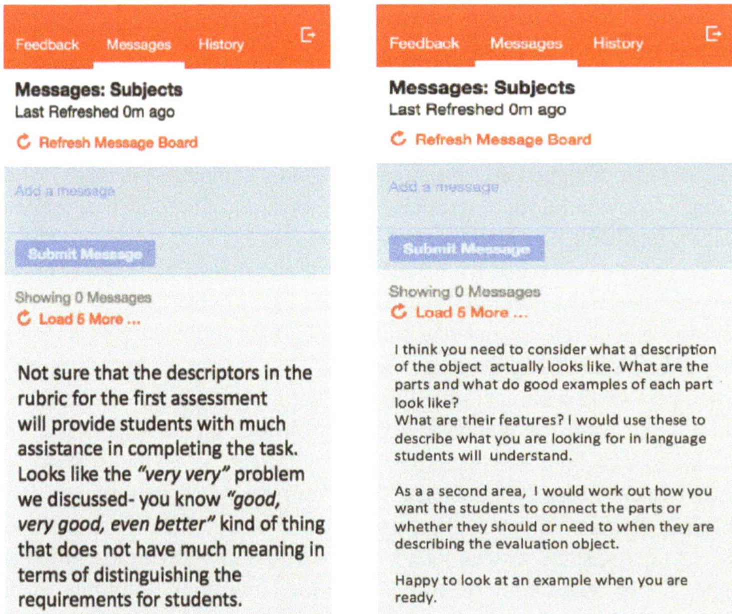

Fig. 4.2 Peer feedback for Jane's assessment task

Second, the revision of Jane's assessment emerges from the agency of her peers along with her commitment to employ their feedback. The feedback and the revision process contributes to the agency of all involved. Jane's revised exemplar will inform those who gave the feedback while the feedback is contributing to Jane's capacity to develop quality assessment tasks. The agency of those involved is an expression of the broader Grandbridge SOU learning and teaching context and its embedded design. The capacity to provide emergent feedback does not rest singularly on attending a professional development program about criterion-based assessment. Because the work of Jane and her colleagues exists within a broader understanding of the learning and teaching context expressed by the university's embedded design, the feedback expresses the community's shared schema for learning and teaching. As an expression of a shared schema, there is a higher likelihood that the feedback will be understood and acted upon. Emergent feedback evolves the agency of those involved, the design process and the schema overall. This stands in contrast to the uncertain position of moderators in the pre-contextual example of 4.1 where no one really knows or feels confident about the feedback being shared and what to do about it.

Third, Jane's work and that of her colleagues is transparent and collaborative. Jane is not working in a private practice context where her work will be judged later in an after-the-fact moderation process. Anyone involved in developing the course or program can use the software tools to look at the ongoing work of peers at any time. The opportunity to share feedback is open and collaborative available constantly as part of the design process. Helping with the design is part of her colleague's responsibility.

Fourth, Jane's work is enabled by technologies. We saw this more expansively in Chap. 3 where the role of technology in course design was described in Example 3.1.[2] What we see here is an extension of that work where feedback is available on a shared platform designed for the timely ongoing exchange of ideas. The technology creates the conditions for collaborative work and the tools that make timely collaborative feedback possible.

Fifth, the data that emerged from Jane's work and that of her team is the richest, most complex, and most important data available about what the university does. It represents the core learning and teaching work of the university at its most fundamental, granular, and sophisticated level. If, for example, the efforts of many faculty members yield a comparable difficulty with rubric design, then that issue or need will emerge from the feedback derived from the shared experience of those working day-to-day on assessment task development. It is the scaled up (less magnified) cycle of timely exchange from normal work that makes feedback emergent. The responsibility of all groups at Grandbridge SOU is to act upon that emergent data in the same way as Jane's team is responding to an identified need in one course. Those teams may be responsible for identifying solutions of broader scope although their work is similar at scale. There is no sense that the leadership dreams up solutions for others to implement. The work of the whole university is a scaled up iteration of the work of Jane's team.

Sixth, because Jane is seeking feedback as she designs her assessment task, she can avert many of the problems that may turn up in a pre-contextual moderation process by improving the quality of an assessment task before it is deployed with students. Her agency and that of her peers is brought to the task just-in-time and before students experience a less well-developed rubric. Moreover, when this feedback is occurring all-of-the-time across a university, problem-solving at scale also becomes an all-of-the time and just-in-time process as feedback emerges at scale from all the work of design, delivery and student engagement.

Seventh, and related to the previous characteristic, the purpose of feedback is focused on dynamic problem-solving and improvement. It does not get bound up in summative judgments about the moderation scores for a single rubric or placing a numerical indicator on the quality of the university's assessments overall (i.e., frequencies of yes and no responses to get a whole of university metric). Instead the focus is on what to do next to improve any and all assessments. There is no difference in response irrespective of the scale at which the problem is identified.

Finally, the feedback shared reflects the Grandbridge SOU learning and teaching context. It is an expression of the university's embedded design for learning and teaching. The feedback instantiates, amplifies, or dampens (Johnson 2001) the attributable relationship among research, practice standards, and the workable distinctions in the quality of what the community does. Emergent feedback makes a

[2]The role of technology in the SOU is taken up in detail in Chap. 7.

proximal contribution to Grandbridge SOU's determination and understanding of productivity because the university knows what it is doing at scale and can make genuine workable distinctions when it comes to the quality of learning and teaching.

4.5 An Emergent Feedback Organization

The self-similar nature of the core work at all levels in a self-organizing university makes possible the conceptualization of an Emergent Feedback Organization (EFO). An EFO is characterized by five core characteristics. An EFO:

1. Is professionally controlled by comparable and visible learning and teaching practice at scale (Bowker and Star 2000). Its learning and teaching form and order flows from its professionally controlled function;
2. Targets agency in its design—through an interaction of people, process, and technology (Latour 2005). Agency is translated across these dimensions and is self-similar (Bigum 2000; Latour 2005). Everything that happens in the organization seems 'personal'/proximal, to everybody and everything in it;
3. Disperses control (Holland 1999) and distributes leadership functions in a dynamic emergent network;
4. Is open and *as* responsive to external forces (external research and trends are part of everyday business) as the feedback it generates from within, and
5. Employs emergent feedback as the way the organization talks to itself (Pascale et al. 2001). Feedback emerges from the agency of those involved at all levels of the organization. Function emerges and re-emerges continuously from the organization's professional circumstances (Bain and Drengenberg 2016).

In Example 4.2, we saw several of the EFO characteristics in action including the role of professionally controlled practice and the agency of the team. Example 4.3 derived from earlier work on learning analytics (Bain and Drengenberg 2016), describes the role of leadership in the SOU using emergent feedback in ways that are self-similar to a course design team in order to highlight the features of the EFO.

Example 4.3 Leading with Emergent Feedback[3]

The Vice President-Academic (VPA) at Grandbridge SOU is preparing for the monthly meeting of the university's Learning and Teaching Team. The team includes the Deans, a program director and faculty member from each of the faculties, and the Manager of the

[3]The scenario in this example is adapted from a similar account of emergent feedback in an EFO in Bain and Drengenberg (2016) *Transforming the measurement of learning and teaching in higher education.*

Office of Governance (manages the university profile of programs). At the meeting the Deans will use the Universities Learning Analytics Tools[4] to report upon the Grandbridge SOU's efforts in the design, delivery and student engagement with assessment. To do so they display an emergent feedback dashboard that presents the faculty profile in the areas of design, enactment and engagement, including student performance. The tools describe real time feedback on the use of pedagogy, quality of assessment and learning experience design, the integration of program standards and student performance on assessment tasks.

The VPA is especially interested in progress with assessment design which is a key element in the university's forthcoming accreditation review as well as it's current strategic plan. Most specifically she is looking at the extent to which rubrics adequately differentiate levels of performance for students. She looks at each of the faculties' reports and notes some differences. The Faculty of Arts is reporting high levels of implementation integrity in the design of their assessment tasks under the new assessment policy. Student feedback on those tasks confirms the quality as do the students' grades. Education and Science seem to be lagging somewhat or experiencing some design challenges. The VPA looks specifically at both high and low performing assessment tasks in the course design software and specifically the way the Arts faculty are designing and differentiating assessment criteria. She makes a note to ask the Dean of Arts to select some positive exemplars to present at the meeting in order to share with the other Deans and to talk with the program directors and faculty members at the group about how they could engage in ongoing cross-faculty professional development work.

THE VPA takes a specific look at the design work being undertaken in a new Bachelor of Medicine degree. The entire team developing the degree, including its leadership, is new to the University and the SOU approach. While the program is yet to be implemented, there are some exceedingly low ratings and lots of gaps in the design data and commentary. She looks at some specific examples to understand the issues. The VPA notes that it may be good for the medicine team to work with some more mature design teams in the University to improve its process and expertise.

We can employ Example 4.3 to reflect upon and clarify the characteristics of an EFO. First, the VPA, the deans and everyone else at the meeting are using the characteristics of professionally controlled practice to make determinations of progress with and the quality of the assessment tasks under discussion. Further, those judgments are based on the ability to link assessment design to feedback on student outcomes. The judgements being made about progress and quality are a scaled up expression of the same criteria being used by Jane as she revised her assessment task in Example 4.2.

Second, Jane's agency, that of her colleagues, the Deans and the VPA are self-similar. While Jane and her colleagues are looking at a course, the University Learning and Teaching Team are looking at the same features for many courses. In many ways, emergent feedback at Grandbridge SOU functions like a GPS system.

We have all had the seamless experience of using a GPS system to guide us to a precise location in a city or locale previously unknown to us. This happens because data, technology, and navigation process including constant emergent feedback combine with the act of driving to lead us to a destination. The GPS system is not a distal representation. It is a real-time transactional guide that works interactively with the dynamic process of making

[4]These tools are described in greater detail in Chap. 7 and reflect the analytic capability derived from faculty using the design tools described in Chap. 2 and in this chapter.

> the journey. We do not have to pull the car over and work out the implications of the symbols on the map, and then stop again to check back and triangulate after we have made some progress, or have someone model and report the results of using the GPS data back to us. (Bain and Drengenberg 2016, p. 54)

We view the GPS journey as analogous to Jane's work as she revises her assessment task and receives feedback from her peers. The protocols, tools, and feedback are all integrated into a navigational process that guides Jane and her colleagues to engage in building courses and specifically assessment tasks. Feedback (data) are indistinguishable from process, and the process is indistinguishable from the tools employed to navigate or transact a course of action—to determine what to do next. This means that the VPA is not simply looking at data, she can see the assessment process. Like a GPS, the process can be viewed at different levels of detail, at the level of a learning experience all of the way to the performance of the university as illustrated in Example 4.3. The process and data does not get more sophisticated by getting bigger if it is aggregated or abstracted in a high level report.[5]

Third, control for the quality of assessment at Grandbridge SOU is dispersed. In preparation for her meeting, the VPA can examine the data and process at any level from Jane's assessment task to the performance of assessment rubrics across the university. She will use the feedback to develop questions, hypotheses and possible actions to be discussed at the meeting, just as Jane's colleagues stimulated the same process with their feedback. The deans' reports are also a self-similar expression of the work Jane, her colleagues and many other faculty members are engaged in across the university's network of program and course designers. The work of the VPA is made possible by the work of those directly involved in learning and teaching who provide emergent feedback.

The shared understanding of learning and teaching makes it much more likely that the members of the University Learning and Teaching Team, or any other team will be able to work together in order to get things done. The model does not diminish the VPA's or any other leader's position in the university. The VPA will not be questioning the deans about execution or acquittal on scores related to the top-down strategic planning objectives described in Chap. 2, rather she will be listening to the deans describe progress on the quality of assessment derived from an established model of learning and teaching practice based upon Grandbridge SOU's embedded design. Everyone's participation references back to an understanding of the learning context as opposed to the connoisseurship described in Example 4.1.

Fourth, the VPA and Grandbridge SOU more broadly are responsive to external forces, including the forthcoming accreditation review noted in the example. The VPA knows the expectations of that review will echo the effort-chain

---

[5]We will discuss the GPS metaphor by way of example in Chap. 7 when we describe in more detail the role of technology in the SOU.

conceptualization of standards, practice and outcomes that represent the accrediting body's current condition and the pre-contextual universities it serves.

This effort-chain accreditation paradigm is a condition or set of circumstances from which Grandbridge SOU has emerged. As a consequence, the accreditation process will not be without its challenges given the order of magnitude of the change at Grandbridge SOU and its adoption of a model that in many respects questions the underpinnings of the current accreditation approach. However, that fact that Grandbridge SOU is enacting in valid and reliable ways much of what is intended or impersonated in the effort-chain accreditation approach creates an important point of connection for the way the university can and will be represented in the review. Grandbridge SOU will be able to show how the university's design for learning and teaching produces the evidence to meet the accreditation standards at scale across the community.

Fifth, Example 4.3 shows the way an EFO can talk to itself (Pascale et al. 2001) because of its understanding of the learning and teaching context and the way the context is expressed in its embedded design. The use of emergent feedback at many levels in a learning and teaching network creates a constant focus on Grandbridge SOU's schema for learning and teaching practice. Jane revises her rubric and receives feedback in a pattern of exchange that is replicated across the university in decisions by faculty members about pedagogy, curriculum alignment, assessment, teaching practice, and student engagement. All of those exchanges generate emergent feedback that is, and will be, the subject of the University Learning and Teaching Team meeting described in Example 4.3 and future meetings. The self-similarity in the focus of assessment work at all levels of the university maintains attention on both the form and function of learning and teaching at Grandbridge SOU. The strategic direction of the university is no more or less than action on and about that emergent feedback.

4.6 Implications

What makes emergent feedback at Grandbridge SOU different from its pre-contextual iteration is that feedback is a genuine part of normal work happening all of the time at the level of interaction among learners and teachers as courses and programs are designed and delivered and as students engage with them. Those interactions can be viewed at many levels from discrete learning experiences to the whole university's performance with assessment although they are essentially the same (Bain and Drengenberg 2016). As noted earlier in the analysis of Example 4.3, only the magnification or perspective changes. A leadership team looking at an institutional report on the quality of course design is looking at the same things a design team is looking at when building a course or program only the magnification changes. By way of comparison and as we have seen in the preceding chapters, data that is poor in conception will not be improved by gathering more of it.

The implications for the leadership of universities of emergent feedback and similarity at scale are immense. Everyone at Grandbridge SOU at all levels of leadership needs to know about and be able to problem-solve and act upon learning and teaching needs and problems in self-similar ways. This does not seem to be an unrealistic expectation of a learning and teaching organization although as we have witnessed it cannot be assumed at any level in the pre-contextual university. In this sense, the SOU constitutes a significant departure from the current modus operandi in the leadership of most universities where conventional wisdom seeks to make sharp distinctions between that which is viewed as strategy creation and the roles and responsibilities associated with executing strategy. We saw this kind of role differentiation at play in Chap. 2 where President Harper saw a clear distinction between his role as a developer of strategy and the operational dimensions of the Pre-contextual Grandbridge strategic plan. The strategy/execution distinction is a longstanding and widely held ethos in higher education despite research largely from the business world that calls into question the validity of such a distinction. According to Martin (2015):

> Execution writers fall into this trap because they want to make a distinction between strategy as deciding what to do and execution as doing the thing that strategists decided. But that begs the thorny question, especially in the modern knowledge economy, what exactly does that "doing" activity entail? If it is to be distinct from the antecedent activity of "deciding what to do," then we pretty much have to eliminate the "deciding" part from "doing." So to create the desired distinction, we would have to define execution as choice-less doing. There are no choices to be made: just do it (whatever "it" happens to be). (p. 3.)

In support of this argument, Martin (2016) cites research showing that studies of leadership effectiveness fail to make a clear distinction between execution and strategy in practice. Leadership characteristics for strategy and execution co-vary —"It's impossible to have a good strategy poorly executed. That's because execution actually is strategy—trying to separate the two only leads to confusion" (p. 1).

We contend that the strategy/execution problem is amplified in higher education because the effort chain approach described in Chap. 1 is incapable of attaining the whole-of organization impact expected of university strategic plans. The underwhelming capacity of the pre-contextual university to exert an influence on learning and teaching at scale creates a basic incompatibility between intent and capacity to execute including the limited capability for generating emergent feedback. This incompatibility encourages leaders to bundle the real work of leading learning and teaching at scale into a basket of activity for downstream accountability called "execution." The latter turns out to be someone else's responsibility somewhere lower on the organizational chart and separate and distinct from high level strategic leadership. From our perspective, the antecedents to this problem lie in the leaders not knowing what to do about learning and teaching at scale in universities, expressed specifically in the lack of understanding of context and the inability to put in place the kind of embedded design for learning and teaching required to produce emergent feedback that exerts a whole-of-organization effect. We will make this case more definitively in Chap. 6.

In the SOU, dispersed control, self-similarity in role along with emergent feedback changes how universities are organized and led. These features make implausible any intent to create a divide between strategy and execution. The role of leadership is an emergent and self-similar expression of the complex and sophisticated interaction between learners and teachers and the feedback generated by that interaction. There is no sense that the feedback generated at that most granular level somehow gets different, more sophisticated as it is viewed at scale; or that considering such feedback in some abstracted form yields greater or different insights that would produce a separate category of work for leaders called strategy. The self-similarity in role and connection to feedback at all levels make any attempt to separate the work of learning and teaching into strategy and execution counterproductive to the overall work of a learning community.

In Chap. 3, we represented embedded design as the architectural step that develops the systems, methods and tools required to translate a university's conceptual understanding of its learning and teaching context into a process for normal work. Emergent feedback as noted earlier in this chapter is the fuel for that working process. It affirms the attributable relationships between practice and student outcomes, establishes workable distinctions in the quality of practice, connects those distinctions to student outcomes and establishes what those outcomes are. It confirms (or disconfirms) standards and enables the dynamic evolution of a schema for learning and teaching at organizational scale. Emergent feedback instantiates the understanding of the learning context that enables a university to show as opposed to claim its effect on student learning. The affirmation of this causal connection means the effects of inputs on outputs in a productivity model can be affirmed based on an understanding of learning and teaching quality.[6] Most important and as emphasized throughout this chapter emergent feedback makes cause and effect connections, establishes quality and makes the legitimate measurement of productivity possible as part of normal work and not as a research study, program evaluation, or acquittal against strategic planning targets.

Takeaways

In this chapter, we described emergent feedback and its role in the SOU. The key takeaway propositions from the chapter are:

- Emergent feedback is the data that emerges from the day-to-day normal professional work of learning and teaching in design, enactment and engagement (Bain 2007; Bain and Drengenberg 2016);
  Emergent feedback is made possible by the way the SOU articulates the learning and teaching context through its embedded design;
- Pre-contextual universities lack the professional control over the learning context to generate emergent feedback at scale;

[6] The implications of understanding quality and productivity are taken up in Chap. 8.

- It is the tight association between feedback and normal work that distinguishes the emergent feedback construct from notions of formative or summative evaluation (Scriven 1967) or evaluation more broadly;
- Emergent feedback happens all the time in a manner that does not exist for time-dependent formative or summative evaluation decisions about the success of a program;
- Feedback that emerges from the fundamental interaction between teachers and learners in design, delivery, and enactment is the most complex and sophisticated data available in a self-organizing university;
- Poorly conceived data that lacks validity does not get better or more informative by getting bigger;
- Examining emergent feedback at different levels of granularity is like using a GPS where the data are viewed with more or less magnification.
- The work of supporting learning and teaching at all levels of the university is similar at scale in the SOU. The self-similar approach stands in contrast to, and is incompatible with, a strategy and execution binary in the leadership of the pre-contextual university.
- Emergent feedback drives constant change in the schema of the self-organizing university as more or less effective practice is amplified or dampened in response to the feedback (Johnson 2001).

References

Bain, A., & Drengenberg, N. (2016). *Transforming the measurement of learning and teaching in higher education*. New York: Routledge.

Bain, A. (2007). *The self-organizing school. Next generation comprehensive school reforms*. Lanham, MD: Rowman & Littlefield.

Bigum, C. (2000). Actor-network theory and online university teaching: translation versus diffusion'. In B. A. Knight & L. Rowan (Eds.), *Researching futures oriented pedagogy* (pp. 7–22). Teneriffe, QLD: Post Pressed.

Bowker, G., & Star, S. (2000). *Classification and its consequences*. Cambridge, Massachusetts: The MIT Press.

Holland, J. (1999). *Emergence: From chaos to order*. Reading, Mass: Perseus Books.

Johnson, S. (2001). *Emergence: The connected lives of ants, brains, cities, and software*. New York: Scribner.

Latour, B. (2005). *Reassembling the Social. An Introduction to Actor-Network Theory*. Oxford: Oxford University Press.

Mandelbrot, B. (1982). *The fractal geometry of nature*. San Francisco: WH Freeman and Company.

Martin, R. (2016, January 12). Strategy and execution are the same thing. *Harvard Business Review*, 1–5. Retrieved from https://hbr.org/2016/01/strategy-and-execution-are-the-same-thing.

Martin, R. (2015, March 13). Stop distinguishing between strategy and execution. *Harvard Business Review*, 1–5. Retrieved from https://hbr.org/2015/03/stop-distinguishing-between-execution-and-strategy

Pascale, R. T., Millemann, M., & Gioja, L. (2001). *The new laws of nature and the new laws of business*. New York: Crown Publishing Group.
Scriven, M. (1967). The methodology of evaluation. In R. E. Stake (Ed.), *Curriculum evaluation* (American Educational Research Association Monograph Series on Evaluation, No. 1, pp. 39–83). Chicago: Rand McNally.

Chapter 5
Governing for Learning and Teaching

5.1 Introduction and Overview

In this chapter, we explain what governance in an SOU looks like including the principles and practices that underpin the approach. In the SOU, governance is an emergent expression of a university's learning and teaching context, its model and professional practice reflected in day-to-day normal work. We have reserved a discussion of the organizational design of governance for the fifth chapter of the book to show how governance in the SOU expresses many of the ideas, methods, and tools described in the preceding chapters.

Most of the literature on organizational design pertains to the field of business and the study of companies. According to McKinsey, while organizational redesign is a high frequency event in the business world, less than 25% of organizational redesign efforts are deemed successful while 40% simply run out of steam (Aronowitz et al. 2015). These finding should be cautionary for organizational redesign efforts in higher education given that businesses can reasonably claim a much clearer and more complete understanding of their professional contexts than universities. Businesses have clearer processes, methods, and strategic intents. They are aware of attributable relationships in the production and sales cycles, receive powerful feedback in the form of sales data, and have a more operational understanding of efficiency, quality and productivity as a result. This does not mean that businesses are invulnerable to dynamic change in products and markets, in a degradation of their own processes, and in changes in competition. The difference is, businesses and other fields that possess a core understanding of their professional practice begin with a deeper and more complete understanding of what they are doing.

In the preceding chapters, we have shown that this relative clarity of purpose and process is not present in higher education where the core activity of learning and teaching is both ill-defined and poorly understood at scale. This ambiguity confounds redesign efforts driving change *around* learning and teaching in ways that are not much *about* it. Most often the result is a costly and disruptive rearrangement

A. Bain and L. Zundans-Fraser, *The Self-organizing University*,
DOI 10.1007/978-981-10-4917-0_5

of the effort chain organizational chart with little demonstrable effect on the quality of learning and teaching. Example 5.1 describes this approach at Pre-contextual Grandbridge University.

Example 5.1 Pre-contextual Grandbridge Redesigns for Learning and Teaching

For President Harper reforming Pre-contextual Grandbridge is a multi-faceted effort that includes generating greater efficiency in the learning and teaching operation. Harper believes that the university faculties are inhibited in their capacity to realize the goals of the university's learning and teaching strategic plan because of the way they are organized. During the past 15 years, the organizational structure of Pre-contextual Grandbridge's faculties have evolved idiosyncratically in response to the needs and challenges presented by their respective programs and the proclivities of the Deans that have led them. Over time they have come to operate as discrete entities. There are significant differences across the faculties in staffing numbers, roles, and methods as they organize, manage and support learning, teaching and the programs they offer.

Harper employs a well-known consulting firm to conduct an efficiency audit of Pre-contextual Grandbridge's organizational design. The consulting team conduct interviews and review the university's policies and procedures. Harper focuses the consultants' attention on the faculty of science which has a preferred model for learning and teaching. The science faculty languished at the bottom of the university's rankings for student satisfaction and instructional quality and has initiated a program called the Collaborative Learning Quality Initiative (CLQI) that has two major components. The first is to make three new appointments to support faculty in the development and delivery of their programs and courses. The new positions include specialists in the following areas: Assessment and Standards Mapping, Advanced Learning Design, and Learning and Teaching Quality. The second component involves a reorganization of those administrative staff supporting learning and teaching to better develop quality control and timeliness. While the effects of the Science faculty approach are unknown there is strong support and lobbying from the Dean of Science and the Vice President (Academic) (VPA). President Harper has also become a strong advocate.

The consultants produce a final report that includes the following recommendations:

1. That the university initiate a common approach to the administrative support of learning and teaching across faculties based on the CLQI.
2. Position the CLQI model within the Center for Excellence in Learning and Teaching (CELT) including three CLQI appointments for each faculty to assure efficiency, and productivity.
3. That services to support students be consolidated at the whole of university level in a Student Support Centre including programs for students at-risk based on data derived from the e-learning system.
4. That Pre-contextual Grandbridge review its committee structure and process for purpose, collaboration, and efficiency.

Harper knows the three CLQI positions will add cost although a number of faculties currently have existing positions that are loosely analogous to the roles identified in the CLQI and will be replaced when the new approach comes under the control of the CELT. He believes the CLQI model will assist all faculties to contribute to the goals in the learning and teaching strategic plan to improve Pre-contextual Grandbridge's national ranking in teaching excellence, student satisfaction, faculty capacity, and award and grant winners.

He is confident that by centralising student support services, Pre-contextual Grandbridge can maximize the use of the analytics capability of the new learning management system while faculties will be freed to pay more attention to the curricular and professional dimensions of learning and teaching. He expects that administrative support will become more efficient and consistent with a model that works across all faculties and includes a reduction in the number of administrative staff overall.

Harper wastes no time in setting up an implementation task force led by the VPA to design a structure that along with the assistance of the consultants will progress the recommendations. The VPA establishes a steering group process to build out the implementation approach for each of the recommendations. Pre-contextual Grandbridge is on the move.

Major organizational reforms in higher education settings frequently follow the Pre-contextual Grandbridge example. Beeson (2014) describes the process as "bringing in a slew of consultants to oversee a large-scale organizational restructuring, most often intended to take out big chunks of cost during an economic downturn (p. 1)." We can unpack Example 5.1 using the three foci most commonly identified in organizational redesign efforts. They are according to Burton et al. (2011), strategy and structure, process and people, and coordination and control. These foci can be viewed within the broader context of Conway's Law (Conway 1968) and the way it plays out at Pre-contextual Grandbridge. The latter means that like an organization's strategic plan, the biggest source of influence on an organizational redesign and change at Pre-contextual Grandbridge will be the way the organization is constituted as it enters the redesign process.

5.2 Strategy and Structure

The most compelling example of the Conway effect at Pre-contextual Grandbridge pertains to strategy. In Chap. 2 we built a case that showed the way Pre-contextual Grandbridge's strategic plan was symmetrical for its effort chain approach. According to Alfred Chandler, based on his 1960s study of four leading US corporations and supported through subsequent decades of study, structure follows strategy in organizational development and further an organization's administrative history is essential to understand its organizational evolution (Chandler 1962). The Pre-contextual Grandbridge experience follows Chandler's edict perfectly. What we see in Example 5.1 is the way the symmetry between strategy and structure is extended to Pre-contextual Grandbridge's organizational redesign. The Pre-contextual Grandbridge strategy, despite its intention to influence the whole organization, relates to making improvements in its effort chain approach, higher rankings, more award winners, more grants all developed in an elective, selective, and exemplary way. The relationship of these things to efficiency, quality and productivity is unknown because Pre-contextual Grandbridge does not know what learning and teaching means at the organizational level, making it extremely difficult if not impossible to take up ideas of efficiency, quality, and productivity in the ways they are addressed in other fields. As we noted in Chap. 1, this difficulty is well established in the literature on measuring productivity in higher education (e.g., Sullivan et al. 2012). None of the strategic planning goals or their implementation steps solve this problem. They are in fact surrogates; they fill the vacuum created by the lack of understanding of the learning and teaching context. They are also incapable of giving any sense of direction to an organizational redesign capable of whole-of-organization improvement.

For example, the expansion of the CELT will add a new line of expertise to its organizational chart enabling the center to grow as an island of learning and teaching excellence. However, it has limited capability (irrespective of mandate) to create whole-of-organizational change at Pre-contextual Grandbridge because there is no shared understanding of the learning and teaching context required to drive a coherent whole-of-organizational change agenda. The CELT will simply get better at making the effort chain work through more grants, more awards, and more professional development. The effect will be to take responsibility away from the organization as a whole and vest it in the expanded capability of the CELT.

The consultants have also attached their agenda and recommendations to what they believe to be an obvious organizational issue: the inefficiency represented by differences in approach among the faculties. The company's consultants know no more about professionally controlled learning and teaching practice than the Pre-contextual Grandbridge leadership. Taking up the inconsistency across faculties is a perfect target for them as that agenda lines up perfectly with the expectations of the Pre-contextual Grandbridge leadership, their (the consultants) capabilities, and avoids the difficult question of what should be happening in terms of learning and teaching at scale across the university. The inconsistencies across faculties are assumed to be problematic although no one really knows about the relationship between the support of learning and teaching and their quality and productivity at Pre-contextual Grandbridge or anywhere else (Drengenberg and Bain 2016).

The models that currently exist in the faculties are not interrogated because there is no way to make a legitimate causal connection between them and each faculty's learning and teaching performance in terms of efficiency, quality, and productivity. In the absence of any evidence, the decision-making process is dominated by connoisseurship-type judgments that privilege the views of key stakeholders including the President. The lack of evidence about the models makes it easy for a new preferred model of support for learning and teaching (the CLQI) to be positioned as a solution because there is no evidence to the contrary and most of the stakeholders do not know what such evidence would look like. Not surprisingly given the extensive lobbying, the CLQI approach is taken up by the consultants whose recommendations centralize the approach in the CELT.

Just as the CELT becomes a larger island of perceived excellence, services to students are centralized and made distinct from the learning and teaching work of the faculties. We see in this recommendation, the way distal data from the new learning management system proposed in the Pre-contextual Grandbridge strategic learning and teaching plan come to define student need as described in Chap. 2. Defining student need with the learning management system data will legitimize and instantiate the separation of students from learning and teaching because the data gathered and used to make decisions has so little to do with learning and teaching interactions. The data is symmetrical for the proposed change in the organizational design. Structure follows strategy in a way that separates teachers from students and learning from teaching.

5.3 People and Process

In the Pre-contextual Grandbridge redesign people and entities become the organizational process. Because there is no whole-of-organization understanding of learning and teaching there can be no whole-of-organization responsibility for it. The new roles, the expansion of CELT, the faculty level redesign, each involve partitioning and segmenting responsibility for learning and teaching in a box and line expansion and realignment of the organizational chart. The flow of information and the way feedback is shared by individuals and entities across the organization does not figure in the Pre-contextual Grandbridge plan because there is no shared professional practice at the organizational level to generate such learning and teaching information and feedback. The CLQI mappers will check that professional standards are covered by programs. Design experts will offer expert advice on learning experiences based on their own autonomously constructed expert practice. Quality assurers will check courseware before it goes live. There will be lots of new checklists and pro formas to complete. Students will attend the Support Centre.

The new experts and expanded entities will assume responsibility for learning and teaching to whom faculty members then become accountable. Because there is no shared schema and practice, roles intended to serve the faculty members function in the inverse. Faculty members simply get more work to do as they become subject to new requirements and accountabilities because the new entities and roles generate a need to be served. Some faculty members will attend professional development sessions conducted by the CELT. Others will not. This will stimulate high-level discussions about making professional development compulsory or finding ways to deliver content more easily online. Instead of a shared and collaborative enterprise, learning and teaching become a set of requirements, things that faculty members need to do in order to fulfill expectations and meet new quality standards.

The review of committees at Pre-contextual Grandbridge will focus on a structure that determines whether those individuals and entities are doing their job, or managing risk, and on ex post facto checking up on the checklists and pro formas after faculty members do their work. When problems do emerge new sub-committees and task groups will be created to deal with the exceptional nature of the new issues. The sub-committees and task groups will replicate the same distal methodologies as their parent groups. There will be clear evidence of due diligence but few workable solutions. An addiction to the organizational chart as a proxy for understanding the learning and teaching context plays out in predictable ways. There will be more structure although no more positive outcomes.

5.4 Coordination and Control

Because individuals and entities become proxies for an understanding of context and organizational process at Pre-contextual Grandbridge, control becomes centralized in those who lead and manage those individuals and entities. The consultants' recommendations position those individuals and entities to assume even greater primacy in the redesign as the entities get more centralized and bigger as a result. Those with aspirations for greater influence become strong advocates of the changes. Despite the rhetoric of collaboration and support in the titles of initiatives and recommendations at Pre-contextual Grandbridge, control cannot be dispersed and leadership distributed because of the lack of a visible and comparable understanding of what is being led and managed.

Successful organizational design should leverage the capabilities of people working in the organization (Price et al. 2009). However, this is only possible when we understand what those people do. There is no doubt that all of the new positions and entities will have role statements that include bullet point descriptors about responsibility for best practice, assessment, leadership, etc. Individuals will be hired who fulfill the requirements of the descriptions although these descriptors have limited meaning because Pre-contextual Grandbridge has no comparable and visible practice at scale to give meaning to the bullet points. Terms like quality assessment, pedagogical leadership, etc., are analogous to the standards described in Chap. 1. They are not understood at a whole-of-organization level. Even the people on the hiring committees at Pre-contextual Grandbridge will not know, in any professionally controlled sense, what those bullet points mean. The successful applicants will take up their positions and invent their own autonomously constructed personal model of Pre-contextual Grandbridge learning and teaching because the university does not have one. They will then report to other people higher up the organizational chart who have responsibility for them and the entities within which they work (Bain and Drengenberg 2016). Those people will use their individually and autonomously constructed personal schema for the Pre-contextual Grandbridge model to lead and judge those they supervise.

The black-box reorganization of learning and teaching at Pre-contextual Grandbridge is unable to produce the workable distinctions in the quality and productivity of normal work necessary to lead and manage. Power is centralized with the connoisseurs, the mappers, the experts, and quality controllers whom along with those who lead them assume responsibility for what goes on. Coordination and control are exercised from top to bottom in ways that are directed at the community of learners and teachers not with them.

The climate and culture at Pre-contextual Grandbridge are not based on a shared understanding of practice, quality, and excellence because there is none beyond the correlates in the effort chain. Climate and culture are dependent upon the community's perceptions about those who lead and their position on the organizational chart absent any empirical understanding of what the organization does and how well it is doing beyond making progress reports on the effort chain goals in the strategic plan and compiling the results of student satisfaction measures. Power, control, coordination, climate, and culture are all person-centric, a logical consequence of the lack of a shared schema, process, and understanding of the learning context and an organizational design incapable of meaningful improvement in learning and teaching at scale.

Pre-contextual Grandbridge will busy itself with the recommendations, setting up sub-committees and task groups, developing relationship maps and flow diagrams, monitoring activity and reporting, ensuring community participation and communication, managing risk, and addressing impact. The wider community will become preoccupied with compliance and the compelling immediate concerns like job security, job reassignment, new processes, and role modifications associated with what will be a significant organizational change process. The inevitability of the change, busyness, and the immediate stressors at Pre-contextual Grandbridge will cause the restructure to morph from an effort intended to improve the efficiency, quality, and productivity of learning and teaching to one focused on the immediate requirements of getting the new structures in place. There will be limited time and opportunity for feedback. The resultant lack of response will be interpreted as agreement and support for the redesign. The difference between making a structure to support learning and teaching and just making a structure will be lost in translation, not that anyone would really be able to detect a difference.

There is no indication that the redesign is or will be informed by preexisting professional knowledge of organizational design beyond what seems to work best for the effort chain. Consistent with Conway's law and Chandler's "*structure follows strategy*" edict, Pre-contextual Grandbridge will opt for a centralized and segmented divisional[1] structure in pursuit of whole-of-organization improvement. The Pre-contextual Grandbridge strategy informs its organizational design in a way that is completely symmetrical for its effort chain approach. The connection with learning and teaching efficiency, quality, and productivity will be assumed although never tested. Creating a common structure to support learning and teaching, the expansion of the CELT and the Student Support Centre, have become convenient proxies for genuine strategic guidance about the way learning and teaching at Pre-contextual Grandbridge is organized. There is no doubt however that Pre-contextual Grandbridge's structure is following its strategy. Key performance indicators may be met although it is unlikely that those accomplishments will improve learning and teaching at scale.

5.5 The SOU Governance Design

In the preceding chapters, we have shown the way terms like professional practice and emergent feedback all mean something specific and operational in the SOU. They are elements in the learning and teaching context and define the scope of

[1]Common approach to organizational design described by Burton et al. (2011).

normal work. For example, in the SOU professional practice is not an abstract idea, bullet point on a position description, or mission-level goal; it pertains to comparable and visible learning and teaching approaches like cooperative learning and active lectures embedded in technologies that are used day-to-day for the design and enactment of learning and teaching work. These approaches have discernable features that produce workable distinctions in the way they are used. Similarly, emergent feedback pertains to the specific way the members of teams respond to each other's work in blogs, and in formal feedback tools using the SOU technology. Features that normally exist as global statements of purpose or intent have meaning in the SOU. This operational quality sets in place the architecture and platform that are keys to the SOU organizational design.

An SOU is comprised of a network of learning and teaching teams. All teams operate with the same simple rules or commitments. They are:

1. To implement the model of learning and professional practice;
2. To use emergent feedback to problem-solve learning and teaching needs;
3. To contribute to the ongoing evolution of the model and schema by sharing and acting upon feedback across the community.

Each learning and teaching team irrespective of its role in the university is self-similar, focusing on the same things as all other teams. This means the executive leadership team at Grandbridge SOU functions with the same three rules as a program team developing a new degree or a learning experience for a course. Figure 5.1 describes that network.

Three key ideas are represented by the network diagram. First, there is no hierarchical structure that creates a distinction or difference in role. All teams are focused on the work of learning and teaching irrespective of their location in the network. The only difference is in perspective and the way perspective influences

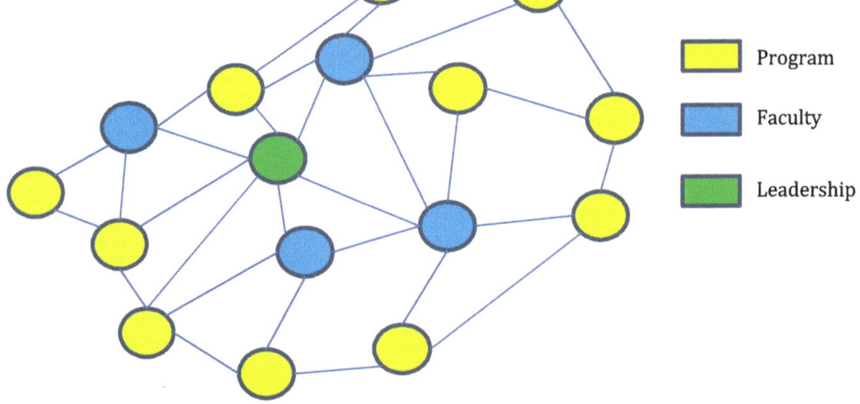

Fig. 5.1 The SOU network of teams

the teams' scope of work in a relatively flat network structure. By perspective we mean the elevation at which the team performs its role. For example, a program team will fulfill its role for a program, the faculty for the profile of degree programs it manages, and the leadership team for the whole university. All teams are interdependent each relying on the work of the others. Their operation is analogous to using a GPS system as first described in Chap. 4 where a user can look at a map from the perspective of a house or a street all the way to the state or country in which the house or street is located. Anyone clicking on the map is using the same data, the only difference is the amount of data used more or less to provide a particular perspective. We use the terms *house*, *street*, *suburb* as designations. They are basically descriptors or waypoints on a continua of perspective on the same thing—a location.

In the SOU, the leadership team may be responsible for looking at the way needs and successes manifest university-wide, although its problem-solving and sharing will involve accessing data at "*street level*" in courses or programs. Similarly, a faculty team may want to share approaches and see what is occurring in programs and courses in other faculties (*suburbs*) or conversely to view a whole-of-university perspective (*state or country*) as it fulfills its role. The same could apply to a program team building a new degree. In line with the GPS analogy, nodes (the circles in the diagram) at the program level in Fig. 5.1 show a direct connection to the university leadership as well as other teams. The network is free of hierarchy. Information from discrete learning experiences within courses all the way to whole-of-university performance are available to all of the teams all of the time as they share feedback and problem-solve.

Second, the program level connections are not restricted to the faculty in which a program resides. Because Grandbridge SOU has a shared learning and teaching model and context, there is a rationale for sharing and exchange across faculties. This is an especially important and distinctive feature of the SOU. A course team struggling with a course design issue in an Accounting degree can find solutions from a team in a different faculty responsible for a Dentistry degree because they are working in the same learning and teaching context with the same whole-of-organization comparable and visible model for the conduct of their normal work. This is depicted in Fig. 5.1 by the connection between nodes at the program and faculty level. Faculties do not operate as silos as was the case in Pre-contextual Grandbridge.

Third, although not immediately obvious from the figure, the membership of teams is in part self-similar. Members of program teams serve on faculty teams while faculty team members serve on the leadership team. All of the teams are connected by their self-similar roles, shared membership, and access to the same data using the university's feedback tools. These shared features shorten the edges (lines joining nodes) in the network making it possible for example, for the vice-president (academic) to have a detailed knowledge of the programs in the university in ways that are the antithesis of the strategy/operation divide that characterizes the hierarchical structure of Pre-contextual Grandbridge. Example 5.2

describes how the network of teams function using the three simple rules that guide their activity.

Example 5.2 The "Very Very" Problem in Assessment Scales Up

Jane serves as a faculty member representative on the faculty learning and teaching team and in that capacity she is involved in the review of other courses and programs in the faculty's profile. In her capacity she is interested to see whether the very, very problem, the inability to adequately differentiate levels of performance on the rubrics (first described in Chap. 4) is more widespread. As she reviews the analytics tools at all levels and programs across the faculty she finds a number of courses with rubrics that in one way or another lack specificity. She also recognizes that in the early stages of a new process where members of the community are building capacity with a new professional skill that these weaknesses may not always be detected in the design feedback as faculty members build expertise with a new approach. She decides to use the Grandbridge SOU feedback tools to check whether there has been any change in student performance across the university since the introduction of the CBA approach (Fig. 5.2).

Jane notices that, as expected the distribution of grades has become more negatively skewed (i.e., more higher grades and fewer low ones) and attenuated with the introduction of CBA and that there are more students achieving grades at the higher level. However, the distribution has changed little with full implementation of the CBA approach in year 2 of the implementation. Jane decides to investigate further within her faculty. She identifies three programs with strong CBA rubrics in their programs and three with weaker rubrics and looks at the grade distributions described in Fig. 5.3.

Jane finds that students in those programs with better CBA are performing at higher levels. She recognizes the risk of inferring cause from correlation from her preliminary analysis. The differences could be related to a range of factors in the design and delivery of the programs sampled. However, those factors are also inclusive of Grandbridge SOU's learning and teaching context and can be interrogated in the same manner as the CBA issue. Jane is on track to find and solve a problem that exists for her course, the program, and possibly across the faculty and university. She raises the issue and shares the data and her experience at a forthcoming faculty learning and teaching meeting. The faculty team agrees to examine the data and the performance of students. Further, the Dean who also serves on the university leadership team agrees to take up the issue from a whole of university

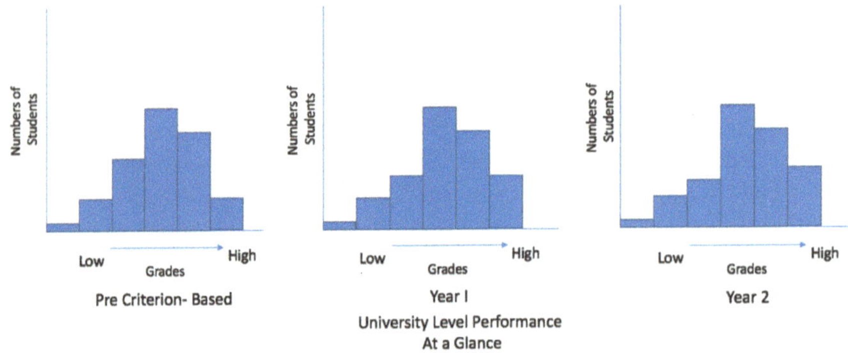

Fig. 5.2 University-level grade distribution on all graduate courses

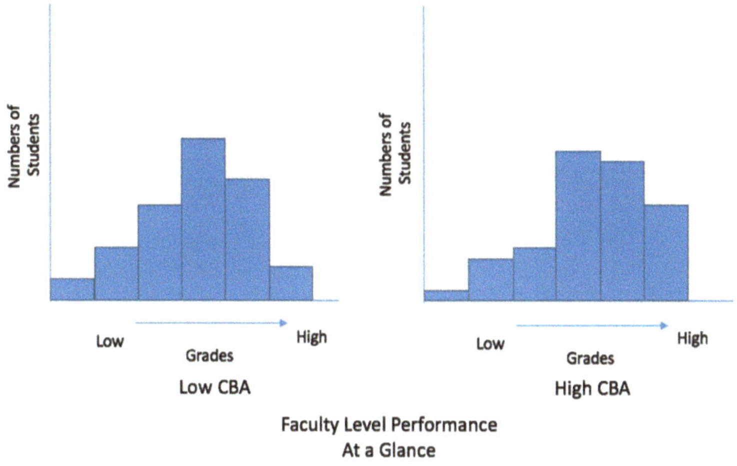

Fig. 5.3 Faculty-level grade distribution on low- and high-quality CBA programs

perspective and repeat Jane's interrogation of the learning analytics data across faculties to see whether the problem exists more broadly. Jane's initial view is that much more work needs to be done on CBA and specifically on the development of rubric criteria that assist students to understand their assessment task requirements.

We can unpack Example 5.2 from two perspectives: First, in terms of the way a network of teams functions using the three simple rules for teams in the SOU. Second, according to the broader characteristics of organizational design used to reflect on Example 5.1. They were structure and strategy, people and process, and power and control.

The example shows learning and teaching teams actively taking up the first simple rule—implementing the Grandbridge SOU model and more specifically the new CBA assessment approach. As we saw in Chap. 4, Jane identified a problem in the same way an educator in any university could employ the interaction with students and their work product to identify and analyze a difficulty with learning and teaching. What is different for Jane is the way she can move forward with her findings within a learning and teaching context that is understood at the level of the organization. She employs feedback from colleagues who are equally engaged with the model and its application in the broader Grandbridge SOU learning and teaching context. The data Jane reviews in Example 5.2 reflects the implementation of the model across the university. For example, the blog posting by Jane's colleague (i.e., Fig. 4.2) indicates an understanding of CBA and the characteristics of assessment rubrics that is currently building across the community. While Jane missed the contribution, it remains available to her for future action. Using feedback that emerges from the student assessments, the course design, and enactment process, Jane and the learning and teaching teams can interrogate the issue from multiple perspectives and points of elevation (i.e., course program, faculty, etc.). Importantly, Jane can build attributable connections between student performance and assessment based on what is occurring in real time at Grandbridge SOU. This is possible because of a shared schema and context that includes a known attributable relationship between quality assessment and student learning. It also includes the network of learning and teaching teams and the technologies that constitute the organizational design at Grandbridge SOU. The teams on which she and others serve are venues for implementing the Grandbridge SOU learning and teaching model.

In the example, we see Jane actively engaged in problem-solving (the second simple rule) as she draws more people and teams into the process. The feedback first identified in the design of her course is linked to other sources of emergent feedback including student performance. As she looks at other programs and courses and especially those with strong CBA, it is likely that she will find solutions to her underdeveloped rubric in the work of her peers. Jane and the teams on which she serves can make attributable connections between design, enactment, and student performance using the tools at her disposal. Learning and teaching are made transparent as well as being visible and comparable using the tools. As she seeks to compare programs with stronger and weaker CBA, she also views the problem from multiple points of elevation. Her agency ultimately helps the whole organization do better with CBA as the work is progressed by the Dean at the faculty and university levels.

In all instances, the feedback Jane uses is emergent, resulting from the day-to-day normal work of designing and enacting her learning experiences. The flatness and short edges in the Grandbridge SOU network of teams supported by Grandbridge SOU's analytics technology, enables the Dean to engage in the same self-similar problem-solving analysis as Jane using the same data to look at assessment practice across programs. These efforts are not separate projects for working parties or sub-committees. They are what everybody does all of the time. Like the GPS analogy described earlier, only the elevation or perspective changes. All teams satisfy the second simple rule in self-similar ways. Because feedback and problem-solving are emergent and exert an influence at all elevations in the university from course to faculty and to the whole organization, the solutions that emerge from the problem-solving effort are ultimately expressed at a whole-of-organization scale. The teams are not waiting to find that faculty members are not filling out checklists before acting on a problem or need.

A solution can only be valid and relevant if there is a shared understanding of the problem or question—a shared schema. For example, solving the "very very" problem is of little consequence to someone who does not know or use CBA. Any solutions that may emerge from the work of teams described in Example 5.2 are an emergent expression of a need that is shared and understood across the community because it emerged from the normal visible and comparable day-to-day work of learning and teaching. The teams described in the example work with the same data in self-similar ways because the community understands and works with a shared schema in a known learning and teaching context.

The shared need for solutions builds collaboration and a more sophisticated representation of problems (Newell 2016; Stoyanova and Kommers 2002). The solution to the "very very" problem at Grandbridge SOU evolves the schema by lending greater depth and meaning to the criteria developed for CBA tasks and in doing so makes the university's teachers more responsive to the needs of their students. The process described in Example 5.2 when applied to all aspects of learning and teaching at a whole-of-organization level will make Grandbridge SOU more adaptive and responsive to changing circumstances and needs. It will also produce a distinctive approach to learning and teaching especially when compared to competitors deeply engaged in developing and instantiating their own versions of the effort chain approach.

In the SOU, problems with learning and teaching are being addressed all the time by the whole community as part of normal work. Issues and challenges are not sequestered as someone else's problem. They do not fall to an assessment mapper, quality controller, or sub-committee to solve. Making things better is the day-to-day work of faculty members as they design and enact their programs and courses and determine the performance of students. Any sense of scaling up emerges from this day-to-day normal work. Such work evolves the schema making Grandbridge SOU more adaptive to the needs of its students.

5.6 Dynamic Strategy = Dynamic Structure

As we described in Chap. 2, strategic planning in a pre-contextual university is a time-dependent stock take of the current reality that inscribes a change in direction or alternative course of action through a goal setting and implementation process. The goals and process are then transmitted to the organization with a possible outcome being a change in organizational structure required to address the goals.

In an SOU, strategy is emergent and dynamic. Organizational adaptation occurs all the time because the kind of activity described in Example 5.2 emerges constantly in all areas and for all aspects of learning and teaching. Strategy means dynamic adaptation to changing circumstances. From the perspective of the SOU, waiting for the big event, the forum, or the five-year plan to invent the future of an organization is to court failure because the circumstances that drive organizations to change are dynamic and ongoing and will not wait for a strategic planning cycle or brainstorming retreat. As Jane reaches out to colleagues and other teams, her work and that of her colleagues at Grandbridge SOU build more edges between the nodes on the Grandbridge SOU network. The richness of these connections reflects the capacity of the organization to adapt dynamically and strategically to its needs and challenges. Structure emerges from strategy.

The idea of strategy as an event or a plan that in turn determines structure is for organizations that work *around* what they do and have not built the systems, methods, and tools to understand their context and generate the emergent feedback that drives dynamic change in their core activity. In the SOU it is altogether possible to have an annual retreat or strategy meeting because it is the beginning or end of the year, because an external report is due, or just because it is a good time to bring people together to discuss emergent feedback. However, such events are discretionary waypoints, occasions when what happens all of the time is examined sometime. It is not a critical summative planning occasion that *pulls the strategic future direction* of the organization out of a *strategic planning hat*, but is instead an opportunity to bring the community together to think about the way the organization is traveling all of the time.

5.7 Where Have All the Committees Gone?

It is remarkable how much work is created in an organization because it does not really know what it is doing. When for example, an organization does not know how to create visible and comparable assessment practice and understand how that work is transacted at scale, it needs committees and working parties to compensate for that lack of understanding. Groups are required to check the checklists that are employed to check on assessment tasks. Because assessment is not understood as a process it is defined as a list of things that need to be done, monitored, risk-managed, and quality-controlled. Under these circumstances compliance is conflated with quality, and risk management with success. The entities created in the line and box organizational structure become the home for the checkers, experts, and quality controllers. Those entities require their own meetings. For example, to be a responsible entity, the CELT will meet and create agendas that mirror the work of the university's higher level leadership group (e.g., developing their own goals and mini strategic plans). The agendas of the entities assume organizational lives of their own in the absence of any processes and feedback that emerges from a known and shared model of whole-of-organization professional practice. Everyone gets busy with activity around the black-box. In doing so, they exercise their fiduciary responsibility for learning and teaching that includes trying to work out what is going on and then make decisions that are transmitted to the community. All of this time and effort is ancillary to normal work because normal work is unknown in a visible and comparable professionally controlled sense.

By way of contrast, because an SOU understands its learning and teaching context monitoring, approving, benchmarking, recognizing excellence, reviewing, and improving are part of the ongoing self-similar work of individuals and the learning and teaching teams using emergent feedback to implement the model, address problems and evolve the schema. In Example 5.2, the review of assessment practice, we saw a way for these responsibilities to be exercised within the scope of normal work across the network of learning and teaching teams. That work did not require separate committees, sub-committees, and task groups to divide and segment into discrete activity that which is simply a part of whole-of-organization learning and teaching. An SOU needs fewer committees, while the teams that do exist are deeply engaged in the sophisticated work of learning and teaching that happens all the time not just when a meeting is called.

It is important to recognize that the scope of learning and teaching organizational committee work in universities extends beyond the example described here. In the

SOU any additional learning and teaching committees function with the same simple rules described for the network of teams described earlier. If, for example, Jane applies for promotion her record of performance will be built from the emergent feedback about student, performance, course and program design, and student feedback; the same data employed day-to-day to fulfill her role as a Grandbridge SOU faculty member. A promotions committee reviewing Jane's application will use the same tools, the same assumptions, and most important the same understanding of the Grandbridge SOU context to determine the quality of Jane's application. Most important, their judgments will not be based on effort chain connoisseurship. They will use quality benchmarks that have emerged from comparable and visible professionally controlled practice at Grandbridge SOU and employed all the time by its teachers and students. Those benchmarks focus on student outcomes and emerge from the feedback shared across the organization as illustrated in Example 5.2.

5.8 People and Process = Collective Intelligence

If feedback in the SOU is emergent, not time-dependent, and mediated by tools that can be used by individuals all of the time, why would we have teams in the first instance and why are dates and times for key organizational activity even necessary? The answer to this question pivots on the idea of collective intelligence and the powerful social nature of problem-solving. Collective intelligence, a term first coined by Douglas Engelbart referred to as collective IQ, meaning a measure of how well people can work together on important challenges. This means how quickly and intelligently they can anticipate or respond to a situation, leveraging their collective perception, memory, insight, vision, planning, reasoning, foresight, and experience into applicable knowledge (Doug Engelbart Institute 2016). Much of Engelbart's work related to augmenting human intellect with technologies to address wicked problems (Engelbart 1962) including the development of hypertext and the computer mouse. The big idea here is that teams are great places for people to share perspectives and insights and pool the collective capacities associated with their membership.

What we also know about the work of teams is that to function effectively teams need a shared understanding of context and a shared language to communicate their views and leverage individual capability for collaborative benefit. The collaboration literature is replete with advice that indicates the need for shared mental models (Fransen et al. 2011) and professional language (Bain et al. 2009). The SOU provides this common shared platform or foundation for collaborative work. Unlike the committees of Pre-contextual Grandbridge where people bring their autonomously constructed and individual schemata to distal effort chain committee meetings, collaborative teams are venues for problem-solving because they are part of the whole-of-organization learning and teaching context. The work of teams at Grandbridge SOU can occur in real-time team meetings or informal exchange among team members and through asynchronous use of the tools, or in combination.

Collaboration in the SOU is not groupthink, the excessive concurrence that occurs when individuals value the group and membership over the decisions it makes (Janis 1972). Collaboration is a robust process that works for better outcomes, challenges incomplete ideas, and builds them into mutually acceptable solutions. It employs well-established collaborative meeting process to enable stakeholders to clarify problems and needs, identify alternative solutions, and work to implement and then evaluate demonstrable courses of action (e.g., Friend and Cook 2014).

The record of emergent feedback captured in the university's feedback tools is the fuel for collaborative team process. Value is assigned to feedback based upon its utility in the problem-solving process. Input that produces professionally informed solutions is amplified because such input is capable of producing workable distinctions within the SOU context. In Example 5.2, we saw that better CBA rubrics covaried with improved student performance. Further interrogation would yield an attributable relationship between the assessment approach and student learning or not. As Jane worked with her team to build solutions the improvement of her rubric and those of her colleagues would emerge from the collaborative process.

Emergent feedback is not filtered, edited, or assigned differential value. Weak feedback is dampened because it is not used or useful in solving problems, recognizing excellence, addressing needs, or producing workable distinctions in the learning and teaching context. High-quality feedback is amplified because of its utility (Johnson 2001). All feedback is welcome.

Further, the role of the individual remains paramount in the SOU. As we saw in Chap. 4 the record of individual contribution in the provision of emergent feedback is central to leadership and recognition in the SOU where faculty members help themselves by contributing to the learning and teaching community. Jane's initiative and problem-solving with others will improve her course and program as well as other courses and programs across the university. Her record of innovative solutions and collaborative action is part of the emergent feedback process and her professional growth at Grandbridge SOU. Instead of building a case for promotion based upon effort chain activity that is at best loosely correlated with actual learning and teaching capability, the cumulative record of Jane's contribution to learning and teaching in generating solutions, providing feedback, and working as a team member at Grandbridge SOU will be the centerpiece of her progression through the learning and teaching career path. She will help herself by helping others and the university overall. We will visit this process in detail in Chap. 6.

5.9 Dispersed Power and Control

In this chapter, we have intentionally juxtaposed two examples that on face value seem quite incongruent. One pertains to a major organizational change, the second is about solving a quite specific design problem in a faculty course profile. The first is profoundly high level and strategic; the second profoundly ground level and operational. Our intent in juxtaposing these scenarios is to draw attention to the fundamentally different construction of scale, complexity and agency in the SOU.

According to nineteenth-century sociologist Gabriel Tarde, understanding complexity, difference and ultimately meaning are derived from the study of things in their fundamental form (Tarde 1893). As we saw in Jane's first encounter with the very very problem in Chap. 4, the more you drill down, the more complex, not simpler things become. This stands in contrast to more standard assumptions about decision-making in organizations where high-level thinking using aggregated data and analysis are deemed to be of greatest import as drivers of organizational action (e.g., the Pre-contextual Grandbridge strategic plan). Tarde would assert that to genuinely understand a phenomenon is to study it at its smallest scale and that things always get simpler and often more simplified as the study of a phenomenon scales up or data is aggregated, modeled, and summarized (Bain and Drengenberg 2016).

We contend for example that the likelihood of improved assessment or any other aspect of learning and teaching at Grandbridge SOU will be a function of day-to-day "infinitesimal[2]" study and problem-solving associated with activity that is closest to the interaction of learners and teachers as described in Example 5.2. This data is needed at all levels as illustrated in the examples employed in this chapter. The learning derived from that study and problem-solving when empowered by the SOU learning and teaching network is fundamental to the development of solutions irrespective of the elevation at which those solutions will be applied. It is from the study of multiple small scale efforts (e.g., examining specific assessment practice in other courses and programs) that the scope of any need emerges and is scaled-up through the collective action across the network. The leadership team at Grandbridge SOU cannot function without access to and the use of such ground-level data. The cumbersome and disconnected committee structure and process in Pre-contextual Grandbridge is what happens when decision-making data is inadequate, and the edges connecting nodes in the network are either too long or do not exist at all. The idea of definitive high level decision-making is an illusion produced by a hierarchical organizational structure and insufficient feedback.

The history of change in higher education and the evidence cited throughout this book would suggest that improvement in practice almost never occurs the other way around by driving distal strategic planning goals through the organizational structure to impact day-to-day work at scale. Further, attempting to change an organization in the manner described in Example 5.1 is unlikely to have any substantive effect because it does not address the fundamental distance problem between high-level planning and ground-level action and the centrality and complexity of the latter as the fundamental source of solution generation and adaptation in organizations. This is more than an issue of elevation but a fundamental difference in what can be reasonably accomplished when information, power, and control exist

[2]Term used by Tarde to emphasize the primacy of small scale effort.

in an abstracted context and when that abstraction is privileged as the primary driver of organizational action.

Jane, working in a networked organization that understands its learning and teaching context is much more likely to influence the quality of assessment than President Harper, his leadership team and "slew" of consultants as they reorganize Pre-contextual Grandbridge. In the SOU, power and control are distributed phenomena because the understanding of the learning context represents a platform for the legitimate distribution of professional agency and influence in a network of self-organizing teams.

The evolution of the Grandbridge SOU schema, the adaptability of the university to changing conditions, and most importantly its responsiveness to students are contingent upon the way it is organized for learning and teaching. The agency of individuals and teams within the broader Grandbridge SOU context, the solution generation for challenging problems and the evolution of the Grandbridge SOU schema makes better learning and teaching possible.

Takeaways

In this chapter, we describe a different way of thinking about organizational design and its role in the SOU. The key takeaway propositions from this chapter are

- Structure follows strategy in organizational change;
- In the pre-contextual university people and entities become proxies for the lack of understanding of the learning and teaching context;
- In the pre-contextual university power and leadership are centralized in the people and entities that assume responsibility for learning and teaching;
- The exercise of power is based on connoisseurship because there is no whole-of organization understanding of learning and teaching as a platform for collective intelligence and action;
- The organization of an SOU is based on a flat network of self-similar collaborative teams;
- The learning and teaching model and context provide the platform for team work that focuses on pooling collective intelligence;
- Emergent feedback is always most powerful at its source and closest to the interaction between learners and teachers;
- In an SOU, a shared understanding of the learning and teaching context makes distributed leadership both possible and authentic.

References

Aronowitz, S., De Smet, A., & McGinty, D. (2015). Getting organizational redesign right. *McKinsey Quarterly*. Retrieved from http://www.mckinsey.com/business-functions/organization/our-insights/getting-organizational-redesign-right.

Bain, A., Lancaster, J., & Zundans, L. (2009). Pattern language development in the preparation of inclusive educators. *International Journal of the Teacher and Learning in Higher Education*, *20*(3), 336–349.

Bain, A., & Drengenberg, N. (2016). *Transforming the measurement of learning and teaching in higher education*. New York: Routledge.

Beeson, J. (2014, January). Five questions every leader should ask about organizational design. *Harvard Business Review*. Retrieved from https://hbr.org/2014/01/five-questions-every-leader-should-ask-about-organizational-design.

Burton, R., Obel, B., & DeSanctis, G. (2011). *Organizational design: A step-by-step approach*. Cambridge, UK: Cambridge University Press.

Chandler, A. (1962). *Strategy and structure: Chapters in the history of the industrial enterprise*. Cambridge, MA: M.I.T Press.

Conway, M. E. (1968). How do Committees Invent. *Datamation, 14*(5), 28–31.

Doug Engelbart Institute. (2016). About collective IQ. Retrieved from http://www.dougengelbart.org/about/collective-iq.html.

Drengenberg, N., & Bain, A. (2016). If all you have is a hammer, everything begins to look like a nail—How wicked is the problem of measuring productivity in higher education? *Higher Education Research & Development*. doi:10.1080/07294360.2016.1208640.

Engelbart, D. (1962). Augmenting human intellect: A conceptual framework. Report prepared for Director of Information Sciences Air Force Office of Scientific Research, Washington, DC. Retrieved from http://www.dougengelbart.org/pubs/papers/scanned/Doug_Engelbart-AugmentingHumanIntellect.pdf.

Fransen, J., Kirschner, P. A. & Erkens, G. (2011). Mediating team effectiveness in the context of collaborative learning: The importance of team and task awareness. *Computers in Human Behaviour, 27*, 1103–1113.

Friend, M., & Cook, L. (2014). *Interactions: Collaboration skills for school professionals* (7th ed.). New York: Pearson.

Janis, I. (1972). *Victims of groupthink: A psychological study of foreign-policy decisions and fiascoes*. Boston: Houghton, Mifflin.

Johnson, S. (2001). *Emergence. The connected lives of ants, brains, cities, and software*. New York: Simon and Schuster.

Newell, C. (2016). *Academics' perceptions of collaboration in higher education course design*. Charles Sturt University, Bathurst, NSW: Unpublished master's project.

Price Waterhouse Coopers. (2009). Point of view. Key principles of organizational design: Diagnosing issues in a company's structure. Retrieved from https://www.pwc.com/us/en/people-management/assets/key-principles-organization-design.pdf.

Stoyanova, N., & Kommers, P. (2002). Concept mapping as a medium of shared cognition. *Journal of Interactive Learning Research, 13*(1/2), 111–133. Retrieved from http://www.editlib.org/p/10783/.

Sullivan, T., Mackie, C., Massy, W., & Sinha, E. (2012). *Improving measurement of productivity in higher education. Panel on measuring higher education productivity: Conceptual framework and data needs*. Committee on National Statistics and Board on Testing and Assessment, Division of Behavioral and Social Sciences and Education. Washington, DC: The National Academies Press.

Tarde, G. (1893). *Monadology and sociology*. Melbourne: Re.press.

Zundans-Fraser, L., & Bain, A. (2015). How do institutional practices for course design and review address areas of need in higher education? *Higher Education Research & Development, 35*(4), 841–853. doi:10.1080/07294360.2015.1137883.

Chapter 6
Agency and Leadership for Learning and Teaching

6.1 Introduction and Overview

In this chapter we take up a second inconvenient truth[1] in the modern university: *Despite the fact that the core business of a university is learning and teaching, those who engage with the effort frequently do not know a lot about it in a professional sense.* Much of the activity of Pre-contextual Grandbridge described in the preceding chapters has its antecedents in this problem. We saw the ways in which the development of strategic plans, mission, and value statements along with other effort chain activity occupy the pre-contextual university. We have also shown these initiatives to be approximations of the real work of leading learning and teaching at organizational scale. This circumstance, which is highly similar from university to university exists in large measure because those leading learning and teaching only have an approximate or analogically reasoned[2] understanding of how to build out a valid and reliable learning and teaching approach at scale in their organizations (Bain and Drengenberg 2016). Their agency is an articulation of what they know and can be expressed within and reinforced by the prevailing form or architecture of the universities in which they work. Further, given the widely recognized hierarchical and top-down organizational structure of most modern universities (Bleiklie and Kogan 2007), the impact of those leaders is profound. Evident in this set of circumstances is a further echo of Conway's law, where the activity of the organization reflects its social construction and architecture (Conway 1968).

---

[1]A term popularized by the 2006 motion picture of the same title about Al Gore's environmental campaign. We described the first inconvenient truth in chapter one as the lack of visible and comparable professional practice at scale in universities David et al.(2006).
[2]This refers to the way universities use the existence of award winners, facilities, and grants as a basis for their understanding of what learning and teaching means.

A. Bain and L. Zundans-Fraser, *The Self-organizing University*,
DOI 10.1007/978-981-10-4917-0_6

In the following pages, we frame the issue of agency for learning and teaching as a *wicked problem*[3] for universities and describe how the problem is addressed in the Self-Organizing University (SOU). If, as described in Chap. 2 the cornerstone of quality learning and teaching is visible and comparable professionally controlled professional practice (Bowker and Star 2000), how is this capacity deployed in the SOU approach? Answering this question by focusing on the agency of those who work in universities is the purpose of this chapter. We begin by using a large-scale study of leading learning and teaching to support our contention about the pre-contextual state of agency in the modern university and demonstrate the wickedness of the agency problem. We follow with a second example to explain how this problem plays out at Pre-Contextual Grandbridge University and then conclude the chapter with a description of the promotion process in the SOU approach as a comparative example of agency at Grandbridge SOU.

6.2 What Is Agency?

Agency is a model of action made possible by a particular body of knowledge that is capable of altering outcomes (Hirschmann 2009). Agency reflects and is developed by the context in which it occurs. All the examples describing the work of leaders and faculty at Grandbridge University, whether pre-contextual or in the SOU represent agency. The actors in those examples go about altering their circumstances based upon the context in which they operate and the knowledge they possess about their roles.

According to Latour (1996) agency is not restricted to human activity. Latour coined the term actants to denote human and non-human actors in a network that shape what they do as a result of their interaction. For our purposes, agency and actants are also closely related to the premise of Conway's Law (Conway 1968) and Chandler's structure follows strategy edict (Chandler 1962) from Chap. 5. Both Conway and Chandler amplify the relationship between the non-human architecture of an organization and the behavior of human actors. We saw examples of actants in Chap. 2 where the form and structure of the Learning Management System (LMS) produced data that shaped the meaning of learning and teaching and the agency of teachers and students at Pre-contextual Grandbridge. Technology exerted a form of virtual agency (Levy 2001). In Chap. 5 we described the way in which the structure of an organization, its committees, organizational chart, roles, and process, influence the behavior of leaders and faculty members as they made decisions about the approval of courses and programs. Human and non-human actants interact to produce agency that reshapes and alters the meaning and outcomes of learning and teaching at Grandbridge in both its pre-contextual and SOU forms. This connection

---

[3]A wicked problem is one that is difficult or impossible to solve because of incomplete, contradictory, and changing requirements that are often difficult to recognize (Wikipedia 2016).

between agency and context is critical to our understanding of the problem addressed by the SOU approach as a strategic or design solution to the challenge of getting to scale with quality and distinctive learning and teaching in universities.

6.3 Agency in the Pre-contextual University

Learning Leaders in a Time of Change (LLTC) (Scott et al. 2008) is a large-scale national study of 513 senior learning and teaching leaders from 20 Australian universities. The findings of the study are wide ranging and address the agency of academic leaders from a number of perspectives. They include policy formation, budget, staff, and relationship management; the capabilities that count for effective leadership, including cognitive flexibility, emotional intelligence, and empathy; markers of effective performance including producing quality student outcomes; the implementation of new initiatives, and improving the quality of learning and teaching.

The outcomes of the study (as described by the authors) include the development of a validated capability framework, proven methods for handling key challenges, quality checkpoints for leadership learning programs, and a revised approach for leadership selection and development in universities (Scott et al. 2008). Overall, the LLTC study speaks at some level of scale for the leadership of learning and teaching in Australian universities. In describing the capabilities that matter for effective leadership the study concluded the following:

> the study has demonstrated that effective leaders of learning and teaching in Australian higher education not only possess up-to-date knowledge and skills on the area, they are also self-aware, decisive, committed, able to empathize with and influence a wide diversity of people, are cognitively flexible, and are particularly deft at diagnosis and strategy formation." (Scott et al. 2008, p. xv)

Given the case made in the preceding chapters, an obvious response to the quote and the LLTC study overall might be: How did the authors come to this conclusion without valid and reliable determinants of quality? Further, given the knowledge of what it takes to be an effective leader, which of the leaders are effective and what impact have they had on learning, teaching, and student outcomes in their institutions?

A close examination of the study helps to answer these questions. The LLTC study represents a profound example of the way the pre-contextual social construction and architecture of the field is reflected in the agency of its leaders. Most importantly, the aforementioned quote and the study overall provides insight to the remit of leadership in a pre-contextual field. For example, there are 21 tables in the LLTC report detailing leadership competencies, none of which refer to the kind of visible and comparable professionally controlled practice (Bowker and Star 2000) that has been shown to influence student learning in education and is the hallmark of professional contexts in other fields. The most specific reference to professional practice among the numerous lists of generic leadership skills indicates that leaders need to have "up-to-date knowledge of learning and teaching" or understanding how to develop an "effective higher education learning program". What that up-to-date knowledge means or what constitutes an effective program are not addressed in the study nor is there any indication that such a body of knowledge exists. The latter did not prevent the authors from claiming the competence and capability of the respondents.

None of the major influencers identified by leaders related to competence, knowledge, or expertise with learning and teaching or issues of organizational impact and scale. The LLTC study also reports participants' leadership tactics including "*keeping in mind that context and culture count, remaining calm when things go wrong, being accessible, telling staff what really counts, and being careful when calling meetings*" (p. xvii). Table 7 of the report includes a list of analogies used to describe the nature of the leadership role from the perspective of the respondents. The list included the following: "*like trying to drive a nail into a wall of blanc-mange- little resistance but no result; being the minister of a church where only the converted come; living in a medieval castle, and rowing without an oar.*" (Scott et al. 2008, p. 50) . Notably, many of the analogies cited in Table 7 allude to the frustration experienced by the learning and teaching leaders associated with the lack of scalable professional practice in those environments in which they worked. Despite the presence of these clues in the data, there was no reference to the lack of professional learning and teaching contexts or as previously noted issues associated with professionally controlled practice at scale in the list of major influences on the respondents' work. This issue was not taken up in the authors' analysis of the analogies beyond reporting the high level of frustration expressed by some participants.

It is also both ironical and alarming that the authors chose analogies and tactical anecdotes as the most detailed representation of leadership perspectives. None of the anecdotes or tactics described, implied, or inferred that leadership work was based upon or informed by an understanding of comparable and visible professional practice at scale. Further, none of the products developed in the study reflected such an understanding. Imagine for example, the likelihood that capability frameworks could be developed in fields like medicine, law or engineering based on generic leadership and interpersonal skills, anecdote, and analogies and without direct and specific referencing to the professional practice of those fields. We see in this study the kind of exceptionalist treatment of the problem described in Chap. 1, where the authors account for a long list of anecdotes that show the frustration associated with the lack of professionally controlled practice at scale with a commentary about the need for negotiations skills, discussion of the complexity and humanness of the environments in which the leaders work, and the need for good planning.

One reason for the lack of attention to the professional learning and teaching context for the work of leaders in the LLTC study is clear—any discussion would yield 513 different autonomously constructed definitions and schema that lack any shared foundation of comparable and visible professionally controlled practice.

Terms like quality, effectiveness and capability are used routinely throughout the study in ways that suggest they are empirically derived and universally understood in higher education. For example, achieving quality student outcomes was described as the top marker of leadership effectiveness. However, as we have noted in preceding chapters, there is no body of evidence to support a causal relationship between what universities do at scale in terms of quality learning and teaching and student outcomes necessary to validate claims about the impact and effectiveness of learning and teaching leadership (ACT 2009; Liu 2008; Pascarella and Terenzini 2005). According to Scott and Scott (2015) the very existence of a knowledge base about learning and teaching in general is questioned by many academics at all levels in universities. We also know that extensive research, including the major summative report of the *National Research Council Report, Improving Measurement of Productivity in Higher Education* (described in Chap. 1) indicates that measuring quality in learning and teaching in higher education is problematic, unresolved, and insufficiently robust to be included in measures of learning and teaching productivity (Sullivan et al. 2012). The latter includes the student satisfaction measures that have become the most widely used and accepted metric for measuring teaching quality in universities.

The LLTC study is an echo of the discussion of standards described in Chap. 1. There we indicated that standards for assessment or quality teaching in both compulsory and higher education have limited effect because they lack an underpinning body of professional practice that lends substantive meaning to them. For example, respondents in the LLTC study reported expansively about the management of time, the leadership of innovation, emotional intelligence, and other dimensions of leadership and management unperturbed by the absence of a clear shared understanding of what is being led and managed at scale in any professional sense. The outcomes of the study as a consequence were frameworks, and guidance based upon generic management, cognitive and interpersonal skills, the definition of role by analogy, and problem solving by way of generic heuristic or anecdote. The LLTC study is far from being an outlier in this regard. The leadership literature in education at all levels is replete with self-report and case studies that talk expansively about the roles and responsibilities of leadership in various guises (e.g., distributed, instructional, transactional, autocratic facilitative, etc.) absent any scaled up understanding of the professional practice of learning and teaching at scale.

In essence, the LLTC study is an all too common case of "*we know what it takes to be effective, skilled, decisive and self-aware* etc,— *just ask us,*" whereby the agency of the university learning and teaching leaders involved in the study reflect the prevailing pre-contextual knowledge base and architecture of the organizations in which they work. The results show the way those circumstances are normalized and perpetuated by self-study that reinforces an effort chain formulation including the methods and assumptions employed to study the agency of its leadership.

We introduced the LLTC study as a profound example of the wicked problem of agency in the leadership of learning and teaching in the pre-contextual university. Why? Because the LLTC study shows there is no perceived need on the part of the authors or higher education leaders involved in the study to ask the elephant in the room question about what is being led and managed, or to call into question the empirical foundations for claims about what constitutes effectiveness, skill, or quality associated with leading learning and teaching at scale. There can be nothing more wicked in problem solving than a circumstance whereby those individuals critically positioned to solve a problem do not believe they have one, or understand what the problem is. The LLTC study suggests leaders in the field are quite comfortable defining their agency without knowledge of scalable models of learning and teaching, or in making a determination of their own knowledge, skill and effectiveness under circumstances where there are no empirically derived benchmarks and standards for professionally controlled practice required to determine quality and productivity. The broader higher education community also appears comfortable with this circumstance.

Our purpose in unpacking the LLTC study is not to pillory a piece of well-intended research or to simply create an additional echo of the key theme in the book, but to highlight the wickedness of the problem of agency in the pre-contextual university. While throughout the book we have made reference to "the university" in its SOU or pre-contextual circumstances, the inconvenient truth is that in making such a reference we are really talking about the people who lead and work in those universities, the community that day-to-day, engage in the activity intended to realize the learning and teaching agenda and aspirations of their universities. The LLTC study of one country's 513 leaders of learning and teaching illustrates how leadership is constructed in pre-contextual ways where leading learning and teaching is not much about the professional practice of learning and teaching at scale and where such a circumstance is not seen as much of a problem beyond the expressions of frustration by some participants.

Further, the inability to articulate what learning and teaching means amplifies the role of generic interpersonal skills in learning and teaching leadership and inhibits the capacity to design organizations in ways that give virtual agency to organizational structures, process, and technology tools that can interact powerfully with human activity for better learning and teaching outcomes. This kind of retreat to relationship factor is common in fields that do not possess professional control (Bain and Drengenberg 2016). It should come as no surprise that leaders of learning and teaching in the modern university build their strategic plans; redesign their organizations, and purchase technology systems in ways that mirror and are symmetrical for these pre-contextual circumstances. Following is an account of the way the strategic plan first described in Chap. 2 is playing out at Pre-contextual Grandbridge as it wrestles with the wickedness of its learning and teaching problems.

6.4 Pre-contextual Agency as a Wicked Problem at Grandbridge

Wicked problems have four characteristics (Rittel and Webber 1973). They are based on incomplete or contradictory knowledge and as a result evoke competing perspectives from many stakeholders; they are expensive to solve involving many interconnections with related issues; they are large in scale and the challenge they create is often unprecedented; the answer or solution is rarely simple (Camillus 2008; Rittel and Webber 1973). Example 6.1 describes how the wicked problem of leading learning and teaching at Pre-contextual Grandbridge University plays out in the implementation of a new assessment policy.

Example 6.1 Pre-contextual Grandbridge Receives Some Difficult News

Over 18 months into the implementation of Pre-contextual Grandbridge's strategic plan (described in Chap. 2) President Harper is meeting with his learning and teaching leaders to review the most recent results from both the Grandbridge University Student Satisfaction Survey (GUSSS) and the National Assessment of University Teaching (NAUT); the latter being an annually administered survey of learning and teaching in all universities (first described in Chap. 1). The NAUT makes possible comparisons of student satisfaction among institutions. Both the Grandbridge and national surveys show that students are not very satisfied with the quality of assessment and the feedback they are receiving in their courses. The results show that the university is scoring at 2.8 on a 5-point point scale in the GUSSS (a small change from previous years) and at the bottom of the third quartile of all universities on the NAUT survey (a slight improvement on previous years). These are not the findings required to position Pre-contextual Grandbridge as the learning and teaching leader described in the university strategic plan.

Frustrating for President Harper, is the apparent lack of return on the 3 million dollars the university has spent developing a new policy and practice framework for assessment. The framework includes the new criterion-based assessment (CBA) requirements for all Pre-contextual Grandbridge courses and programs, an expanded professional development program for all faculty (delivered by the upgraded Center for Excellence in Learning and Teaching-CELT) ; the peer moderation process; the development of the new software-form attached to the LMS that enables faculty to peer moderate and quality assure their assessments by checking whether the guidelines for CBA are implemented; the positioning of assessment as a priority in the university's new promotion framework, and a new website also developed by the CELT that includes resources and examples for implementing CBA. President Harper is looking for explanations as to why there has been no substantial change in the response of the student body to Pre-contextual Grandbridge's assessment practice after an institution-wide "full court press" to improve the university's approach. Especially disconcerting is the fact that eliciting better feedback from students about assessment was a key performance indicator on the Pre-contextual Grandbridge strategic plan; a major plank in the reinvigoration of the university's online learning offerings, and efforts to reposition Pre-contextual Grandbridge as a national and international leader in online learning.

The Vice President (Academic) (VPA) indicates that the program has only been in place for 18 months and may not yet be showing the return on the new reorganization (described in Example 5.1), professional development, and moderation approaches. The Director of CELT indicates that more professional development is needed. While most faculty members have completed the *Introduction to Criterion-Based Assessment* program, the Director believes an advanced module is required for all faculty especially if they are to do a better job of linking assessment to instruction. The Director of Information Technology reports that the data from using the software moderation form shows that over 97% of courses have been successfully moderated for better assessment practice although there has been some faculty resistance in using the moderation software form because of the time involved. The latter is seen as intrusive extra "checklist work" by many. The Director of Information Technology believes many of the problems could be solved by refining the online form. The Director of Human Resources indicates that in the last round of promotions there was a 2% increase in successful applications and a number of faculty members included a range of self-identified innovations in portfolio assessment in their applications. The Director is surprised about the lack of progress given what has been a greater focus on assessment in the promotion process. The Deans indicate that all programs have now met the CBA requirements and are unsure why there is so little change in the way students feel about assessment. After extensive discussion, President Harper and the group recognizes that additional work needs to be done to address the issue in pursuit of better outcomes. They decide to develop a supplementary refresh of the action plan to address the situation. They identify the following action items:

1. The CELT to develop and implement a compulsory second professional development module for linking assessment to instruction;
2. The Deans to conduct an audit of assessment tasks in all courses to try and explain the current problem and ensure quality and compliance;
3. To include an assessment requirement in all promotion applications;
4. To refine the online moderation form in order to make it easier to use.

President Harper directs the leadership to double down and implement the refreshed plan immediately with an expectation of an improvement in student feedback over the next twelve months.

We can use the four characteristics of a wicked problem to unpack the difficulties the Pre-contextual Grandbridge learning and teaching leadership is experiencing with its new approach to assessment. First, the knowledge base on which Pre-contextual Grandbridge is building its assessment approach is contradictory and incomplete. While the core research knowledge underpinning CBA is compelling (e.g., Hattie 2008), the existence of that knowledge at Pre-Contextual Grandbridge is limited.

Further, as we described in Chap. 3, the model of providing supplementary professional development in contexts that do not have preexisting professionally controlled practice has not worked in any sector of education (Cole 2012).

There is also an issue of incomplete knowledge and understanding at a more granular level. From President Harper's perspective, the problem exists because students are reporting they do not feel especially positive about what is happening with assessment at Pre-contextual Grandbridge. Those perceptions are derived from two sources of evaluative data—the GUSSS and NAUT student satisfaction measures. Data derived from these measures are included as evidence of outcomes in the Pre-contextual Grandbridge strategic plan. As we noted in Chap. 1 and Example 3.2, student surveys provide a sense of how students feel about their experience in a course. The data derived from the GUSSS and the NAUT are unlikely to provide the leadership with the feedback they need. The Pre-contextual Grandbridge leaders do not know enough about the research to practice underpinnings of the assessment approach; the status of the implementation; what is working well and what is not, and the specific points of breakdown to respond. The latter represents the kind of information required for focused problem solving, solution generation, and action planning. We know according to the moderation analytics, moderation is happening. According to the Deans the policy is being implemented. Everything Pre-contextual Grandbridge intended in its strategic plan has happened yet the expected results are not forthcoming, and the leadership can only speculate about what is going wrong given the lack of detailed emergent feedback about what has actually transpired. Further, given what we know from the findings of the LLTC study and Jane's experience in Chap. 4, it is questionable whether the Pre-Contextual Grandbridge learning and teaching community possesses the knowledge and expertise to act at scale even if the university had more refined and sophisticated feedback about assessment. The action of the leadership is constrained by the incomplete information in its possession and the way Pre-contextual Grandbridge is designed for learning and teaching. They are symmetrical. Better assessment has not been embedded deeply in the way Pre-contextual Grandbridge is designed for learning and teaching. Under these conditions, the community and the leadership specifically defers to what they know as reflected in the "more of the same" effort chain supplementary action plan.

We also see the wickedness of the problem in terms of the different and competing perspectives of Pre-contextual Grandbridge's leaders as they conceptualize the student feedback about assessment. Because Pre-contextual Grandbridge has no shared schema, understanding, or model of professional practice, attribution about the problem reflects the individually constructed schema, perspectives, roles, and priorities of those involved. For example, the VPA thinks there has been insufficient time for the new approaches to work. For the CELT Director, it is a professional development issue. For the Director of IT, it is a technological problem, etc. Each leader mirrors Pre-contextual Grandbridge's box and line design for learning and teaching. There is no sense that the antecedents to the problem could be a bigger and broader design issue, a reflection of the way Pre-contextual Grandbridge is constituted for its learning and teaching work. The leaders define a complex problem based upon their compartmentalized roles in the effort chain and their positions on the organizational chart. The solutions they propose are a list of "hit and hope" activities because there is no robust base of evidence about the problem derived from emergent feedback or a unifying term of reference to define and then address the issue in an integrated manner.

The Pre-contextual Grandbridge leaders have some sense of the interconnectedness that makes the assessment problem wicked although that understanding is constrained by the limits of their individually constructed effort chain schema. They can visualize the pieces in the jigsaw puzzle (e.g., technology, professional development, promotion framework) but there is no deep understanding of each piece and how they fit together because there is no deep understanding of the learning and teaching context and associated professional practice that would imbue the puzzle pieces with meaning. For example, they recognize that CBA is inexorably connected to the design of instruction. They know the promotion framework, (first described in Chap. 1) should be supporting the new assessment process, and they see a role for technology in the form of the moderation checklist. Unfortunately, while these elements are recognized, they are so poorly articulated and loosely coupled (Weick 1976) as to render them ineffective singularly or even when chained together. For example, as we saw in Chap. 4, there is a pressing need to build faculty member capacity with the skills of CBA at Pre-contextual Grandbridge. The promotion framework could function as a powerful driver for building that capacity at scale. However, in its effort chain form, the framework encourages faculty members to submit any assessment approach they believe to be innovative when applying for promotion. Instead of focusing on the expertise required to produce workable distinctions in the quality of CBA, faculty members are able to meet the promotion requirements without any reference to the approach. While CBA is central to Pre-contextual Grandbridge's broader assessment effort, there is no expectation that submissions for promotion demonstrate how a candidate shows evidence of CBA skill development.

The Human Resource Director is perplexed but should not be. With such a diffuse and disconnected expectation for evidence of competence, the promotion framework is not a driver for better CBA at Pre-contextual Grandbridge irrespective of the number of candidates who self-identify and are promoted with skills in assessment.

The role for technology in solving the assessment problem is conceptualized with the same kind of thinking that drove the purchase of the new LMS, whereby a tool designed to automate instruction was expected to produce better learning and teaching. As identified in Chap. 4, the moderation checklist is not designed to incorporate the evidence-based features of CBA or build capacity with the new assessment approach. Instead, it simply allows users to check whether in their view the key steps or milestones for implementing CBA are occurring according to guidelines. Like the LMS, the moderation checklist automates a learning and teaching process without helping users get more skilled in assessment by using a technology. The checklist has agency but it is not in a form that will produce better assessment practice. Further, in the absence of any process to assure the skills and expertise of the peer moderators, the validity of the summative judgments recorded by the moderation form is largely unknown and definitely untested. As we also noted in Chap. 4, we see in this peer approach the same kind of evaluation by consensus and connoisseurship that becomes the default approach when a system cannot produce workable distinctions in professional practice. There is no shared schema for tackling the problem.

The Pre-contextual Grandbridge learning and teaching leaders do not understand the full scope and meaning of each of their dots, how to join them, or even how to decide what is working to produce a better outcome. Their assessment and overall learning and teaching knowledge may be up-to-date in some generic sense just like their counterparts in the LLTC study but it falls far short of the knowledge and skills required to make the connections necessary to address the assessment or any other learning and teaching problem at scale.

We also see in the scenario the cost issues that make a problem wicked. The assessment problem is expensive to solve. Pre-contextual Grandbridge has already spent 3 million dollars getting its new assessment approach in place. This includes the work of university committees, the technology, and faculty professional development. There are also associated costs that include the redevelopment of the promotion framework, the consultants leading the redesign of Pre-contextual Grandbridge described in Chap. 2 and the new LMS etc., that are purportedly about making learning and teaching at Pre-contextual Grandbridge more distinctive and effective. Unfortunately, the expenditures mirror the effort chain—money is expended in categories—technology, professional development, etc., with little interaction and even less likelihood of systemic effect. Because there is no reliable empirical understanding or model of productivity at Pre-contextual Grandbridge based upon professionally controlled practice, there can be little confidence about the effect of the expenditure on assessment or anything else in the realm of learning and teaching. As more money is spent without demonstrable benefit (e.g., the attainment of strategic planning goals), the more pressure is brought to bear on President Harper and his leadership team.

The longer Pre-contextual Grandbridge persists with its effort chain approach the more likely it is that leaders will ultimately conceptualize the problem as a compliance issue whereby faculty members are not doing what they need or are meant to do. This is likely to happen because the way the university has separated its strategy from execution. Pressure will be brought to bear on middle managers for failing to effectively implement the leadership strategy at an operational level in ways that echo the separation between the roles of strategy and operations in Pre-contextual Grandbridge's organizational structure. We see early evidence of this in the audit of programs and courses to be conducted by the Deans, and in the reports of the IT Director about faculty perceptions of busy work. Leadership, instead of employing emergent feedback collaboratively in the flat network structure described in Chap. 5, instantiates the top-bottom divide between strategy and operations that is unlikely to produce positive outcomes for Pre-contextual Grandbridge. It is all the leadership team knows and can do when the effort chain fails to realize their expectations.

President Harper is frustrated given that one of the most important goals of the strategic plan is not being realized despite the expenditure of significant resources in the area. He is especially concerned that students have responded so poorly to the work completed to date and especially given the overall strategic planning effort is designed to improve student outcomes. Pre-contextual Grandbridge is spending a lot of money to realize its strategic planning goals, although there is an emerging sense that it could be a case of good money following bad unless the leadership can find more pointed holistic solutions to their issues and needs. The expenditures will ultimately demand a claim of victory at some point in the future. Although such a claim is unlikely to covary with genuine change in the distinctiveness, quality and student outcomes associated with the effort.

The fact that the assessment issue is symptomatic of Pre-contextual Grandbridge's failure to build a model of professionally controlled practice is not on the radar of the Pre-contextual Grandbridge leadership. Just as it was not seen as a foundational pre-requisite for the LLTC study of 513 university leaders of learning and teaching. To solve the assessment problem, or the technology problem, or the strategic planning problem, Pre-contextual Grandbridge needs to recognize and address the interconnectedness of those elements beyond its effort chain understanding and realize that to solve any of these discrete problems it needs to put in place the foundational organizational design that can solve them all. Without such an approach the wickedness of the Pre-contextual Grandbridge problem will no doubt direct its leaders to feel they "are trying to drive a nail into a wall of blanc-mange or being the minister of a church where only the converted come or rowing without an oar." (Scott et al. 2008, p. 50) as they go about their work.

Of greatest concern from a cost perspective is the preparedness of Pre-contextual Grandbridge and so many universities to continue to expend resources on initiatives and change processes that we know cannot make a difference. It is from this perspective that the problem at Pre-contextual Grandbridge is most wicked and for which the assessment issue is only a symptom. We have seen this as a recurring problem whether in the unrealistic expectations for professional development, the procurement of technologies that will not improve learning and teaching, organizational restructures that re-arrange the deck chairs but do not proximally influence learning and teaching, and promotion frameworks that lack the focus to produce workable distinctions. Most problematic in relation to cost is the way expenditure begets more expenditure without a better outcome. Without an alternative approach and with the knowledge that peer institutions are doing and spending on the same things in the same ways, good money follows bad with little change in quality, productivity and most important student outcomes.

The problem of assessment at Pre-contextual Grandbridge is a problem of agency; a reflection of the university's design for teaching and learning and its failure to build a shared schema that would make interconnections in the approach to capacity building, technology, organizational design and the advancement of faculty among other things. The inability to resolve those connections in efficacious ways makes the problem messier and more wicked as loosely coupled solutions are layered on top of each other in an environment absent any professional control (Bowker and Star 2000) in learning and teaching at scale. The supplementary action plan reflects this failure.

The Pre-contextual Grandbridge supplementary action plan mirrors the agency of those responsible for planning Pre-contextual Grandbridge's future. It may be too big a leap in expertise and skill to expect those learning and teaching leaders at Pre-contextual Grandbridge to come up with the kind of complete professionally controlled alternative design described throughout this book, just as such an expectation would be beyond the expressed capabilities of the learning and teaching leaders surveyed in the LLTC study. However, at a minimum, Pre-contextual Grandbridge's learning and teaching leaders must assume responsibility for persevering with a course of action that is failing in clear sight at their own institution. This is especially the case given their approach has, as we have seen in the research cited throughout the book, no empirical foundation in higher education efficacy research, a documented track record of failure (e.g., standalone professional development), and will not work to improve the quality of learning and teaching at Pre-contextual Grandbridge or anywhere else. At a minimum, their agency should be expressed in an effort to at least ask the elephant in the room question and put on the table evidence that supports the need for a different way forward.

6.5 Agency at Grandbridge SOU

In the example that follows we describe the promotion process for a faculty member at Grandbridge SOU. The example illustrates the way agency is expressed in a networked organizational design that employs emergent feedback in a professionally controlled context. In this chapter, (as we did in Chap. 5) we juxtapose a high level and strategic example with one that shows the ground-level impact of an effective organizational design. Our intent in juxtaposing these scenarios is to again draw attention to the way the SOU is an emergent approach where key decisions, problem solving and adaptations are an expression of a shared schema and agency at ground level in the normal day-to-day work of learning and teaching.

Example 6.2 Jane Applies for Promotion at Grandbridge SOU

We first met Jane in Chap. 4 when she was wrestling with the challenge of moderation at pre-contextual Grandbridge (Example 4.1) and then in Example 4.2 as she wrestled with an assessment problem. Jane is now contemplating an application for promotion to Associate Professor at Grandbridge SOU with a focus on the area of learning and teaching. She is reviewing the promotion process. One of the simple rules or commitments of Grandbridge SOU described in Chap. 2 is: "*Career progression is predicated on the contribution an individual makes to the growth of others in the Grandbridge SOU model of learning and teaching.*" In many ways this statement is not unlike the aspirational value and mission statements included in many pre-contextual strategic learning and teaching plans or promotion frameworks. Jane reviews the tenets underpinning Grandbridge SOU's promotion framework which gives here a sense of what is meant by the commitment to advancement by assisting others and the university. They are:

1. Recognition and progression focus on excellence in practice combined with leadership and innovation in learning and teaching;
2. Excellence in learning and teaching is first and foremost an emergent expression of excellence in the normal day-to-day work of learning and teaching;
3. The demonstration of excellence for the purposes of career progression emerges from the record of a candidates learning and teaching activity and less reliant on an ex-post-facto construction of prior accomplishment;
4. The normal work of learning and teaching at Grandbridge SOU is collaborative; The attainment of higher levels in the Grandbridge SOU career path requires the demonstration of leadership in learning and teaching that impacts Grandbridge SOU's overall performance and capacity at scale and the attainment of excellence by others in their normal work for and with students;

Rank	Learning and Teaching Practice	Design	Leading	Problem-solving
Associate Professor	• Successfully complete three teaching appointments in the SOU Institute; • Developed or reviewed and refined at least one Institute course in the SOU Course Space[1]; • Quality Institute teaching demonstrated in the Course Space Analytics Tools.	• Develop a program in the SOU Course Space; • Demonstrate emergent feedback has been provided and acted upon; • Demonstrate a track record of providing advanced feedback in the SOU Program, Course and Learning spaces for others; • Demonstrate, through Course and Learning Space Analytics the way design feedback has successfully improved teaching and student learning outcomes.	• Lead a program team for at least one complete program design and implementation cycle to demonstrate the following: • University benchmark performance in delivery and student growth; • Develop team member capacity through problem-solving and team leadership as reflected in team member performance and career progression; • Demonstrate program improvement through response to feedback using Program Space Analytics.	• Present an evidence-based case demonstrating a solution to a major institutional learning and teaching problem that has resulted in a change in policy and practice and is validated using the Program Space Learning Analytics Tools.
	1. The course design software tools that embody the Grandbridge SOU model described in chapter x.			

Fig. 6.1 Excerpt from the Grandbridge SOU Promotion Framework

5. Feedback is the key to the success of the Grandbridge SOU approach. The progression of every individual faculty member pivots on generating and sharing feedback with others (collaborative/group and individual), and taking that feedback to continuously improve learning and teaching practice, system and process design.

Jane can see how the tenets fit with much of her day-to-day work at Grandbridge SOU although she needs to understand what they mean in a more operational way and specifically, as they relate to the requirements for progression to the rank of associate professor. She reviews the Grandbridge SOU promotion framework: Fig. 6.1 describes an excerpt from the framework.

The first thing Jane recognizes from reconciling the commitment and tenets with the excerpt from the framework described in Fig. 6.1 is that her application will focus primarily on her record of normal work in the Grandbridge SOU learning and teaching context. She will not have to invent or reinvent herself as a teacher, designer, or leader in a retrospective portfolio process to progress in the career path. For example, the design work and feedback developed as she solved the very very assessment problem (described in Chap. 4) will demonstrate her advanced understanding of CBA and show how it contributed to a university-wide solution. She believes a case study of that work will meet the criterion for problem solving described in Fig. 6.1. The additional criteria described in the figure make clear how her ongoing record of work as a course and program team leader will comprise the bulk of her application in teaching, leadership and design.

Jane's promotion process is a dynamic and emergent expression of her contribution to learning and teaching at Grandbridge SOU over time. The record of her teaching, leadership, design expertise, and support of others are derived from Grandbridge SOU's Program, Course and Learning spaces software.[4] Her case will be built on a longitudinal record of existing analytic data that make workable the distinctions in the quality of her work. The data record is a product of the emergent feedback she has given and received as she has gone about the normal work of teaching, leading, course, design, delivery and evaluation.

[4]This software will be described in Chap. 7.

> Because the software provides data on all learning and teaching activity at Grandbridge SOU, Jane's work can be benchmarked against the performance of the university, its faculties and her peers.
>
> A second major requirement described in Fig. 6.1 is participation as an instructor in Grandbridge's SOU Learning and Teaching Institute where new and continuing faculty build their skills and knowledge in the Grandbridge SOU model. Teaching in the institute enables Jane to demonstrate her advanced skills in all aspects of the model, the teaching practices, use of the technology collaboration and learning and teaching design. In addition, she will be involved in the development, review, and refinement of the institute syllabus demonstrating advanced skills in teaching and design. In her teaching role at the institute she will also contribute to the growth of her colleagues. In all areas, Jane will be required to "walk the talk" of what Grandbridge SOU means by learning and teaching. The explication of the learning and teaching context and Grandbridge SOU will support her in doing so. On the basis of her review of the commitments, tenets and framework, Jane decides to proceed.

How does Jane's engagement with the promotion process relate to Pre-Contextual Grandbridge's whole-of-organization assessment woes and the LLTC study? They reflect two very different permutations of agency in a higher education approach to learning and teaching. To illustrate the difference, we will return to the four characteristics of a wicked problem in order to show how Jane's application reflects a form of agency that makes learning and teaching in universities much less wicked.

First, wicked problems are based on incomplete or contradictory knowledge. We saw in Example 6.1 and the LLTC study how incomplete knowledge of learning and teaching and specifically the inability to establish a body of visible and comparable professionally controlled practice (Bowker and Star 2000) blunts the capacity for planning and problem solving, and results in a diminished understanding of what leading learning and teaching means. In comparison, Jane's promotion process and the criteria she will address (excerpted in Fig. 6.1) are clear. Her agency is expressed in a context where practice is known and the professional lexicon is shared. As with lawyers, doctors, and other professionals, individual genius, mastery, and need are each expressed within a known professional context.

The promotion tenets and criteria are a tangible "pointy end" expression of a whole-of-organization knowledge base and schema made possible by Grandbridge SOU's embedded design. As we saw in Chap. 3, the SOU combines a university's learning and teaching model, policy, promotion frameworks, technologies, etc., to translate commitments into the systems methods and tools for routine normal work. They are actants in a network of collaborative action. Their agency, when empowered by the engagement of the Grandbridge SOU faculty and leadership produce the workable distinctions in learning and teaching practice necessary to validate a promotion framework and process. Jane's application for promotion is a waypoint on a professional journey, an occasion when what she does all of the time is reviewed more specifically at a given point in time to recognize her contribution. The criteria described in Fig. 6.1 provide the terms of reference to enable such recognition. They reflect the knowledge base and schema for learning and teaching shared by the Grandbridge SOU community. The knowledge base is known and while in a dynamic state of adaptation is complete at any point in time and not contradictory.

The second characteristic of a wicked problem is the way it produces different, competing, and frequently incompatible perspectives. We saw this play out in Example 6.1 where the interpretations of the Grandbridge SOU assessment problem depended on the roles of the leaders and the leadership categories they represent. Their supplementary plan was a list of loosely connected initiatives that were not much different from the original strategy. In the SOU approach, competing or challenging perspectives are framed within the context of emergent feedback described in Chap. 4. Jane's application for promotion is predicated on the way her feedback has worked to solve problems, improve designs, and advance her own capacity, that of the Grandbridge SOU community, and the university's schema for learning and teaching. Within the SOU context, Jane can challenge and contradict to exert a specific influence on the way learning and teaching happens. She did this in developing a solution to the very assessment problem. Her feedback is an emergent expression of a shared schema that enables the agency of individuals to be more collaborative than competitive, complementary than idiosyncratic, and for solutions to be more sharply focused on the specific features of a problem. This is the kind of detailed and nuanced feedback that is missing in the Pre-contextual Grandbridge analysis of its assessment problem and also caused the authors of the LLTC study to defer to generic descriptions of best practice, anecdote and analogy in their account of learning and teaching leadership skills in universities.

The third characteristic of a wicked problem refers to the cost of finding solutions and the risk of expending resources in ways that do not produce successful outcomes. For example, Pre-contextual Grandbridge's supplementary action plan applies additional resources to an approach that has failed to produce an adequate return on investment in its original iteration. The Pre-contextual Grandbridge example is echoed in the equivocal findings of research on the relationship between expenditure on higher education and student outcomes (Pike et al. 2011) that shows expenditure only modestly related to self-reported student outcomes. What is clear from research in all education sectors is *how* the money is spent turns out to be the key consideration in understanding the relationship between expenditure and outcomes. Unfortunately, given the pre-contextual circumstances of the modern university our understanding of what to spend the money on for predictable effect is limited. For example, the current best practice model for determining higher education productivity is based on distal inputs of instructor time and outputs of graduation rates (Sullivan et al. 2012). Not surprisingly, the model lacks a detailed understanding of what instructors do with their time or as we have seen throughout the book how and whether the expenditure of time contributes to those student outcomes. We will discuss this further in terms of learning and teaching productivity in Chap. 8.

Effective expenditure of resources on learning and teaching requires an understanding of what constitutes learning and teaching quality and productivity (Massy et al. 2012). Quality can only be asserted when legitimate workable distinctions in learning and teaching practice can be established at a whole-of-organization scale

(Drengenberg and Bain 2016). Because Grandbridge SOU understands in operational terms the complexity and criticality of ground-level normal work as the foundation for its feedback processes, problem solving, and planning it can discern quality and make workable distinctions in routine practice. Jane's application for promotion is based on doing things that are known to work at Grandbridge SOU. Jane and her colleagues whether applying for promotion, convening a meeting of a design team, working on the design of a learning experience, course, or program in the Grandbridge SOU software, are constantly engaged the learning and teaching context and the visible and comparable professionally controlled practice (Bowker and Star 2000) known to exert a demonstrable effect on student outcomes. Jane's application for promotion can be viewed as a tangible and clearly understood expression of agency linked demonstrably to learning and teaching productivity. There is no need to employ years of experience, peer approbation, or defer to the blunt connoisseurship of promotion committees and generic learning and teaching promotion frameworks as proxies to determine whether she is ready to be an associate professor. Grandbridge SOU understands Jane's promotion application in a way that extends far beyond the loose covariance of instructor time to graduands in existing productivity models.

The fourth characteristic of a wicked problem is: The solution is rarely simple. This is certainly true and a challenge for both Pre-contextual Grandbridge and Grandbridge SOU. For example, Jane's application for promotion is just one element of a sophisticated and complex process of organizational design described throughout the book and specifically in Chap. 3 where the embedded design process was explained. Her application is only possible because Grandbridge SOU has employed embedded design to articulate its learning context at whole-of-organization scale.

Even more challenging though are the circumstances facing Pre-Contextual Grandbridge, as it responds to a faltering strategic plan by doing more of the same using its effort chain conceptualization of agency. The key point here is that while the design process and organizational change described in the preceding chapters are challenging and definitely not simple, the approach builds a design capable of getting a university to the starting line in terms of linking its learning and teaching approach to student outcomes. Such a connection is a fundamental pre-requisite for legitimizing any genuine claims of efficacy in curriculum design, quality teaching, promotion, and the expenditure of resources. This connection will never be made at Pre-contextual Grandbridge no matter how much it plans and spends on its effort chain approach.

While the challenge of taking on the wicked problem of learning and teaching at scale in universities is significant, a whole-of-organization solution offers up a unique opportunity to bring transparency, equity, efficiency, quality, productivity and distinctiveness to a university's approach. It portends a new kind of whole-of-organizational agency as people and design interact for better student outcomes. The alternative restricts the university to claims of effort chain effectiveness without ever creating the clear and demonstrable line of sight from their learning and teaching approach to student outcomes.

6.6 Distributing Leadership?

In concluding this chapter, our intent is to highlight a final key distinction between Pre-contextual Grandbridge, the LLTC study, and Jane's promotion process as they relate to agency, order, control and distributed leadership. While much is written and claimed about distributed leadership in higher education (e.g., Corrigan 2013; Jones 2014; Jones et al. 2014), most of this work falls into the same trap that befell the LLTC study, where what is actually being led is not explained or explored at scale. This makes any claims about the efficacy of distributed or any other form of leadership problematic. We contend that an organization can only effectively disperse control at scale for an activity it understands at scale. This was our fundamental concern about the LLTC study—the way leadership expertise could be assumed, and in some cases claimed without an empirical basis or an expression of clear shared knowledge of the activity being led.

In pre-contextual universities, learning and teaching are least understood at the point of interaction between teacher and student. We saw this in Example 6.1 as the leadership team addressed a complex and multifaceted problem with only the judgment call of leaders and student satisfaction data as a basis for its problem solving. The information available to the leadership team reflected their pre-contextual understanding. It did not get any richer or more complete as it was aggregated at the whole-of-organization level because it was poor in its conception.

A university needs to first understand the sophistication and complexity of professionally controlled learning and teaching practice in its most elemental form to build the systems, methods and tools to engage with professional practice at whole-of-organizational scale. The university's strategic understanding of itself emerges from that complexity. We see this in Jane's promotion application and the promotion framework. Grandbridge SOU understands what it is doing as reflected in the professionally controlled, visible and comparable standards for career progression and the tools it employs to establish whether those standards are being met. The promotion standards are not an adjunct to normal work; they express what Jane does all of the time in her role as an educator at Grandbridge SOU. Because Grandbridge SOU can make workable distinctions in relation to the quality of Jane's work, the university does not have to create a separate parallel process to determine her contribution to learning and teaching.

Any organization needs professional agency in its design to disperse control and distribute leadership. The latter is built on an understanding of the context in which leadership occurs (Jones et al. 2014). The SOU model is predicated on an assumption that instantiating the learning and teaching context based upon a ground-level understanding of comparable and visible professionally controlled practice creates the potential for dispersed control and systemic self-organizing agency. For example, Grandbridge SOU has a promotion committee that manages the promotion process although the decisions of that committee are essentially criterion-based because the criteria for promotion are clear, explicit and underpinned by professionally controlled practice. Jane's promotion and that of her peers is transparent, and valid derived from the empirical record of their normal work and an understanding of the complexity, significance and value of that work as it emerges from the interactions among many students and teachers. The way promotion happens at Grandbridge SOU is a reflection of its broader embedded design and the way agency is expressed in the interaction between its design features and the work of the Grandbridge SOU community. We can see in Jane's promotion experience all of the elements of the Grandbridge SOU design—policy, tools, emergent feedback, etc.

The promotions committee is the keeper of the process. Its decisions are distributed and self-organizing. This stands in contrast to the way promotion decisions are made in the pre-contextual university where all too often powerful leaders in the system stand in adjudication of faculty presentations that as previously noted frequently represent a parallel reality to the normal work of the organization. Those committees use pre-contextual frameworks that rely heavily on connoisseurship to make their decisions as a proxy for the lack of understanding of what learning and teaching means. The governance of the promotion process is tightly held by the top-level leaders in the form of extensive regulation, administrative guidelines, and appeal processes that have little impact on quality learning and teaching at scale. They are proxies for the inability to make and recognize workable distinctions in faculty expertise.

By way of contrast, there are no top and bottom levels of agency at Grandbridge SOU. We can view Jane's work and her promotion experience as being analogous to the GPS described in Chaps. 4 and 5. There we described the way a GPS can zoom in on a location making it possible to view the same feature at different levels of granularity, not to be confused with seeing something different. Jane's work at Grandbridge SOU and her engagement with the promotion process is a zoomed in expression of the whole of Grandbridge SOU. In the SOU, distributed leadership is not so much an expression of one level of the organization giving way to others but where leadership itself is an emergent expression of the organization's schema and design in action—its professional agency. It is this similarity at scale that makes Jane's experience an important juxtaposition to the work of the Pre-contextual Grandbridge leadership and the LLTC study because it shows the way control and ideas of leadership can become a genuinely distributed phenomenon when an organization understands what learning and teaching means and where the agency of an individual is self-similar for the functioning of an organization as a whole.

Takeaways

In this chapter, we describe a different way of thinking about agency and its role in the SOU. The key takeaway propositions from this chapter are:

- A knowledge base and understanding of context are essential in understanding agency and for distributing leadership;
- Higher education has spent little time or energy defining its knowledge base or context in the area of learning and teaching;

- The result is a diminished capacity at all levels for the professional practice of learning and teaching in the pre-contextual university;
- Both Pre-contextual and Grandbridge SOU possess agency that reflects the knowledge base and schema that exist in both iterations of the university;
- Agency in the SOU is an expression of a shared schema that is derived from a professional understanding of the learning and teaching context;
- In the SOU agency at scale is an emergent expression of what learners and teachers do in day-to-day interaction. It is most complex and informative at ground level;
- The relationship between expenditure on higher education and student outcomes is poorly understood and symmetrical for the pre-contextual university's contemporary understanding of learning and teaching;
- Professional agency depends upon an empirically derived understanding of quality;
- Quality is a product of workable distinctions;
- Productivity is derived from the quality of workable distinctions;
- The professional agency of the members of a learning community reflects the organization's embedded design.

References

ACT (2009). *ACT report: Voluntary system of accountability learning gains methodology.* Retrieved from https://cp- files.s3.amazonaws.com/22/ACTReport_LearningGains-Methodology.pdf.

Bain, A., & Drengenberg, N. (2016). *Transforming the measurement of learning and teaching in higher education.* New York: Routledge.

Bowker, G., & Star, S. (2000). *Classification and its consequences.* Cambridge, Massachusetts: The MIT Press.

Bleiklie, I., & Kogan, M. (2007). Organization and governance of universities. *Higher Education Policy, 20*(4), 477–493.

Camillus, J. (2008). Strategy as a wicked problem. *Harvard Business Review.* Retrieved from http://www.induscommons.com/files/102770262.pdf.

Chandler, A. (1962). *Strategy and structure: Chapters in the history of the industrial enterprise.* Cambridge, MA: M.I.T Press.

Cole, P. (2012). *Linking effective professional learning with effective teaching practice.* Retrieved from Australian Institute for Teaching and School leadership website: http://www.aitsl.edu.au/docs/default-source/default-document-library/linking_effective_professional_learning_with_effective_teaching_practice_-_cole.

Conway, M. E. (1968). How do committees invent. *Datamation, 14*(5), 28–31.

Corrigan, J. (2013). Distributed leadership: Rhetoric or reality? *Journal of Higher Education Policy and Management, 35*(1), 66–71. doi:10.1080/1360080X.2013.748479.

David, L., Bender, L., & Burns, S. (Producers), & Guggenheim, D. (Director). (2006). *An inconvenient truth* [Motion Picture]. United States: Participant Media.

Drengenberg, N., & Bain, A. (2016). If all you have is a hammer, everything begins to look like a nail—How wicked is the problem of measuring productivity in higher education? *Higher Education Research & Development.* doi:10.1080/07294360.2016.1208640.

Hattie, J. (2008). *Visible learning: A synthesis of over 800 meta-analyses relating to achievement*. New York: Routledge.

Hirschmann, D. (2009, January 19). 4 more definitions of agency. [weblog post. Retrieved from https://asociologist.com/2009/01/19/4-more-definitions-of-agency/.

Jones, S. (2014). Distributed leadership: A critical analysis. *Leadership, 10*(2), 129–141. doi:10.1177/1742715011433525.

Jones, S., Harvey, M., Lefoe, G., & Ryland, K. (2014). Synthesising theory and practice: Distributed leadership in higher education. *Educational Management Administration & Leadership, 45*(2), 603–619. doi:10.1177/1741143213510506.

Latour, B. (1996). On actor-network theory. A few clarifications plus more than a few complications. *Soziale Welt, 47,* 369–381.

Levy, P. (2001). *Cyberculture*. Minneapolis: University of Minnesota Press.

Liu, O. (2008). *Measuring learning outcomes in higher education using the measure of academic proficiency and progress (MAPP).* Retrieved from https://cp-files.s3.amazonaws.com/23/RR-0847MeasuringLearningOutcomesUsingMAPP.pdf.

Massy, W., Sullivan, T., & Mackie, C. (2012). Data needed for improving productivity measurement in higher education. *Research and Practice in Assessment, 7,* 5–15.

Pascarella, E., & Terenzini, P. (2005). *How college affects students (Volume 2): A third decade of research.* San Francisco: Jossey-Bass.

Pike, G., Kuh, G., McCormick, A., Ethington, C., & Smart, J. (2011). If and when money matters: The relationships among educational expenditures, student engagement, and students' learning outcomes. *Research in Higher Education, 52,* 81–106.

Rittel, H. W. J., & Webber, M. M. (1973). Dilemmas in a general theory of planning. *Policy Sciences, 4,* 155–169.

Scott, D., & Scott, S. (2015). Leadership for quality university teaching: How bottom-up academic insights can inform top-down leadership. *Educational Management Administration & Leadership, 44*(3), 511–531. doi:10.1177/1741143214549970.

Scott, G., Coates, H., & Anderson, M. (2008). *Learning leaders in times of change: Academic leadership capabilities for Australian higher education.* Retrieved from http://research.acer.edu.au/cgi/viewcontent.cgi?article=1001&context=higher_education.

Sullivan, T., Mackie, C., Massy, W., & Sinha, E. (2012). *Improving measurement of productivity in higher education. Panel on measuring higher education productivity: Conceptual framework and data needs.* Committee on National Statistics and Board on Testing and Assessment, Division of Behavioral and Social Sciences and Education. Washington, DC: The National Academies Press.

Weick, K. E. (1976). Educational organizations as loosely coupled systems. *Administrative Science Quarterly, 21,* 1–21.

Chapter 7
Technology for Learning and Teaching

7.1 Introduction and Overview

In the preceding chapters, we have shown why a university needs to know what professionally controlled learning and teaching means at a whole-of-organization level if it is to exert an influence on student learning at scale. In this chapter, we demonstrate how and why an effort chain approach in the pre-contextual university limits the role of technology rendering it a distal influence unable to produce the transformational benefits achieved in fields capable of professionally controlled practice (Bowker and Star 2000). In contrast, we describe the role of technology in the SOU and demonstrate how an understanding of the learning and teaching context is used to redefine the contribution technology can make to student learning. This includes an example of the way new approaches to the use of technology can improve the quality of learning design.

7.2 Transformational Technology?

There is much debate about the role of technology in education. While posited as a transformational force (Christensen et al. 2008), the evidence would suggest that huge improvements in access to hardware, software, and network applications over the last 30 years have not exerted a transformational effect on student achievement. Results in higher education are inconclusive at best (Trucano 2005) and show no clear direct effects on achievement outcomes (Youssef and Dahmani 2010). These conclusions are also consistent with findings from the compulsory education sector. Students at all levels of the education system do not appear to be learning more, faster, or more deeply despite transformational changes in the availability and use of technologies in schools and universities (Bain and Weston 2011; Cuban 2003; Guri-Rosenbilt and Gros 2011; Weston and Bain 2010). Further, studies of

A. Bain and L. Zundans-Fraser, *The Self-organizing University*,
DOI 10.1007/978-981-10-4917-0_7

technology-rich deployments (e.g., 1:1 laptop programs) have not shown anything near transformational achievement effects (e.g., Shapley et al. 2009; Silvernail 2007; Weston and Bain 2010; Windschitl and Sahl 2002).

We contend the source of this efficacy problem can be traced back to an effort chain approach and the failure to bring genuine clarity to the learning and teaching context. Technology cannot influence learning and teaching in design, enactment, and engagement if we do not know what those things mean in a professionally controlled scalable sense. This problem relegates the use of technology to managing or augmenting the learning process. This includes scheduling and student record keeping; in an adjunct curricular role for managing goals, standards and outcomes; or in the form of multimedia resources (Coates et al. 2005).

Given the predominance of this kind of peripheral use, it should come as no surprise that we hear the frequent mantra—technology is "just a tool" in education. Even prominent technologists like Bill Gates and leading educators have positioned technology as an ancillary source of educational support, in ways that also imply in some instances a contest with the role of the teacher:

> Technology is just a tool. In terms of getting the kids working together and motivating them, the teacher is most important.—Bill Gates
>
> It is important to remember that educational software, like textbooks, is only one tool in the learning process. Neither can be a substitute for well-trained teachers, leadership, and parental involvement.—Keith Krueger
>
> Any teacher that can be replaced with a computer, deserves to be.—David Thornburg

(Rao 2012, p. 1)

What do these quotes say about how we understand learning, teaching, and technology and why as the quotes imply do we see a need to affirm the teacher's role in ways that amplify the distinction between technology and teacher? At play in these quotes and the "technology as tool" mantra is an echo of Conway's Law (Conway 1968) first described in Chap. 2 whereby the field unable to define its practice at scale employs technology in a manner that reflects its current architecture and social construction. Technology is just a tool because it is only in the realm of automation (i.e., grades, scheduling, curriculum organization and support and learning management) that the field of education understands and possesses clear line of sight to the role and purpose of technology at scale in educational organizations. The use of technology mirrors the organization's schema for learning and teaching.

The need to make the distinction between teacher and technology is also an interesting logical consequence of the nature of work in a field that lacks professionally controlled practice. The preoccupation with affirming and re-affirming the distinctive value and contribution made by teachers in comparison to new technologies reflects a lack of understanding of and insecurity about how teachers influence learning. Without an understanding of the learning and teaching context and those things both technologies and teachers do to influence learning outcomes, the rationale for the contestation or juxtaposition becomes clear. The efforts of both technologies and teachers are stymied because the field lacks the comparable and visible practice (Bowker and Star 2000) necessary to understand their respective and joint contributions to student learning and how those contributions interact. Example 7.1 describes this problem at Pre-contextual Grandbridge University.

Example 7.1 A New Learning Management System at Pre-contextual Grandbridge

After decades using multiple iterations of a locally developed bespoke learning management system (LMS), Pre-contextual Grandbridge is excited about the deployment of its newly purchased LMS with advanced Web 2 capability.[1] The leadership team views the system's implementation across the university as a cornerstone of its learning and teaching strategic plan and Pre-contextual Grandbridge's capacity to meet national standards for quality and accreditation.

The LMS developers claim the system offers better course documentation for accreditation purposes, a more accessible interface for users, better learning analytics including sophisticated tracking of student use, better collaboration tools and classroom chat, along with excellent conferencing capability, virtual breakout facilities and a range of new interactive applications and customization potential. The university has identified a number of key early adopters some of whom are recent Pre-contextual Grandbridge teaching award winners to develop and test exemplary modules in the new system as a way to highlight its functionality. Members of the group of early adopters have been awarded *Relief from Teaching Fellowships* to assist their peers develop similar modules. The new analytics capability of the LMS will be employed to evaluate the modules after they are developed. The Centre for Excellence in Learning and Teaching (CELT) is also responsible for working with the faculties to implement the LMS and has deployed the majority of its staff in faculty and school-level orientation and consultation roles to ensure a smooth transition to the new system.

The LMS initiative will bring Pre-contextual Grandbridge into line with its competitors many of whom are using the same product. There is an expectation from the leadership team that the new LMS will assist the university achieve its learning and teaching strategic goals of improving student evaluations, expanding the number of promotions in the area of learning and teaching, and increasing grant applications and award nominations. For faculty members there is an opportunity to look in detail at student activity on the system linking activity in their courses to student performance.

The university leadership is especially excited about the potential of the new analytics capability of the system to improve retention and optimize costs associated with learning and teaching. For example, the advent of the new LMS and the associated optimization of teaching quality will bring forward plans to expand Pre-contextual Grandbridge's online offerings and create additional blended learning opportunities.

We can employ the description of a learning and teaching context cycle introduced in Chap. 1 to unpack Example 7.1. The cycle includes the existence at organizational scale of the following: a model of professional practice, attribution and efficacy, standards, workable distinctions (Drengenberg and Bain 2016), and emergent feedback (Bain and Drengenberg 2016; Bain 2007).

[1]Web 2.0 is a second iteration of the World Wide Web that focuses on the ability of people to collaborate and share information online (Webopedia, 2014).

Members of the leadership team are excited about the system because it reflects what they believe learning and teaching means. The LMS mirrors the Pre-contextual Grandbridge schema and the agency of the community. The absence of a whole-of-organization model for learning and teaching is not viewed as a problem because the need for such a model is not on the university's radar. There is an expectation that the new features and functionality of the system will exert an attributable effect on the things Pre-contextual Grandbridge values about learning and teaching as reflected in its strategic plan (e.g., award winners, promotion applications etc.). There is no sense that the anticipated causal relationship between the LMS and student learning is a bridge too far because Pre-contextual Grandbridge is already conflating its effort chain approach and strategic planning goals with student learning outcomes and growth. The LMS is simply an important enabler in the overall plan and fully symmetrical with the strategic goals that do not include a model of professional learning and teaching practice.

Pre-contextual Grandbridge is also pinning many of its hopes for better learning and teaching on the functionality of the new LMS (the collaboration tools, new learning analytics capability etc.,) and student learning. However, there is no body of research that establishes an attributable relationship between LMS use and student learning and achievement (Means et al. 2014). A recent large scale K-12 study by Kimmons (2015) found that learning management systems account for just 1–2% of the variance in achievement ratings, a finding that in direction is broadly consistent with those for other learning and teaching technologies including games and simulations, reading and math software, and tutoring systems that have not added significant value to student learning over the existing learning and teaching paradigm (Bain and Drengenberg 2016; Barbour and Reeves 2009; Dynarski et al. 2007; Santoro and Bishop 2010; Smith et al. 2005; Steenbergen-Hu and Cooper 2014). Further, it seems to make little difference whether these applications exist on the desktop, the local server, or in the cloud. Pre-contextual Grandbridge will instead rely on the developers' assertion that simply using the LMS features and functionality will produce successful learning outcomes. However, existing evidence would suggest this is unlikely given there is no known attributable efficacious relationship between the new Pre-contextual Grandbridge LMS and student learning outcomes at scale.

Pre-contextual Grandbridge could leverage the LMS for better learning outcomes. If the university possessed a model of professional practice, it could integrate the functionality of the LMS with professionally controlled practice and possibly develop the LMS technology into something new and distinctive. For example, the new chat room functionality of the LMS could be employed in combination with research-based cooperative learning (CL) to build group learning activities that exert a powerful influence on student learning. The interaction of the chat room functionality and CL would mean that well-known achievement-related features of CL including the way accountability, interdependence, and learning tasks (Slavin 1996) are organized and structured could be leveraged by and with the chat room tools for learning effect. Feedback about chat room use would focus on the key achievement-related features of CL. We know that there is a greater likelihood of successful student outcomes when technologies are placed within a context of professionally controlled practices that have been shown to influence student learning (Bain and Drengenberg 2016; Hattie 2008; Weston and Bain 2014). However, Grandbridge in its pre-contextual state has no way of enacting the research-based knowledge of CL or any other practice at scale.

It is also possible that some of the example modules developed by the Pre-contextual Grandbridge award winners and referred to in Example 7.1 could incorporate research-based practices like CL. However, there is no guarantee this will be the case as Pre-contextual Grandbridge has no professionally controlled standards to determine what exemplary practice means and what will be included or prioritized in the development of the new modules. The teaching relief fellows will be encouraged to meet and collaborate with peer faculty members as part of their fellowship responsibilities. They will self-invent the peer collaboration process focusing on those things each fellow deems to be of value in module design. Faculty members will get involved at their discretion.

The absence of a model to drive efficacious practice and the development of standards means there will be no genuine workable distinctions in the quality of the design work arising from the effort to develop better modules. The work of the fellows and their mentees will be claimed as an educational victory although differences in the quality of the modules will not be discernable in any sophisticated professionally controlled sense. Participating faculty mentees will be recognized for getting involved and the fellows for their good works, although the recognition will not be linked to any clear distinctions in the relative quality of the work completed. The broader community will view the exemplary modules and the process employed to develop them variously as a source of interest, inspiration, aspiration and/or approximation. Some participating faculty members will document their involvement in the fellowship program as evidence for promotion using the Pre-contextual Grandbridge promotion framework. Others will dismiss the whole idea as a niche pursuit of enthusiasts not really applicable to their mainstream work in university teaching.

Not surprisingly, the learning and teaching feedback available to Pre-contextual Grandbridge faculty members from the new LMS will reflect the data the LMS is able to produce—navigation patterns, downloads, uploads, logons, etc. These will be the criteria employed to determine the quality of the new modules and learning and teaching more generally because these are the available data. Faculty members will receive reports summarizing their performance on these key metrics. The university leadership will receive reports about the faculties. Distinctions in patterns of student, faculty member, and overall university performance will become evident from the reports. Analytics data about the LMS–student relationship will be linked causally to student performance and outcomes despite the lack of credible evidence to support such a connection. Feedback from the LMS will be deployed as criteria for awards or promotion applications. Some faculty members will attend professional development sessions and conferences on how to improve engagement with the LMS.

The effort chain will be awash with data although little will change at Pre-contextual Grandbridge in terms of the quality, efficiency, and productivity of learning and teaching because the features of the LMS like the analytics data it produces are not proximally related to student achievement (Bain and Drengenberg 2016). The proximal learning and teaching factors that could improve student performance at Pre-contextual Grandbridge are not within the university's reach or even in its line of sight as an organization. The university is hoping to assume a position of leadership and distinction in the sector although it is using the same LMS product as its peers in ways that are most likely to produce outcomes highly similar to those peer institutions. While promoted as a key tool for innovation, the reality is that the LMS will simply make pre-contextual Grandbridge more like the other pre-contextual universities with which it competes.

Pre-contextual Grandbridge does not possess the professional practice, standards, organizational design, networks, feedback and combination of expectation, and support mechanisms required to bring the work of the relief fellows to scale. Any lack of uptake of the example modules or digression from them as models of best practice will be explained using language that highlights the importance of maintaining maximum faculty member flexibility, and respecting autonomy and openness in the support of the broader learning community. There is little chance that the good works of Pre-contextual Grandbridge's learning and teaching leaders will exert a whole-of-organization influence on the quality of learning and teaching. It is simply too much to expect of the relief fellows when their capacities have not been established or affirmed by genuine standards or their work supported in anything that resembles a professional learning and teaching context. Technology will continue to have little impact on learning and teaching at Pre-contextual Grandbridge. The teaching relief fellowship program will come to a successful conclusion and the university will move on to its next selective, exemplary or elective new thing.

7.3 Technology for a Self-organizing University

The purpose of technology in the SOU approach is to represent and instantiate a university's learning and teaching model in day-to-day normal work, to improve student learning, generate emergent feedback, and build the organization's learning and teaching capacity. This means embedding the model of learning and teaching and professional practice in tools that generate emergent feedback and produce workable distinctions in routine practice at all levels of the university. Technologies possess the capacity to be agents of actual and virtual professional control (Levy 2001) when they function at the nexus of the learning context, known professional practice, the professional agency of the user, and the affordances of the technology. What is created at that nexus is the product of an interaction among context, user, and technology.

7.4 Edge Technology

We describe ICT tools that serve to instantiate the professional context in interactive ways as edge technologies (Bain and Weston 2011). Edge technologies do five things, they:

- enact evidence-based protocols for professionally controlled practice in terms of content and process;
- connect design, enactment and engagement using the network capability of technology to join these nodes and shorten the edges between them. In network theory, an edge is the connection between two nodes on a network (Barabási 2002);
- extend and distribute capacity through a collaborative human–machine relationship (Facer and Sandford 2009). They do so by translating protocols into conceptual and then relational schema for action;
- enable emergent feedback by connecting learners in open collaborative platforms, shortening the cognitive distance in the learning and teaching network, making feedback an emergent expression of the routine work of design, enactment and engagement. They bring teachers and learners together for collaborative development and feedback;
- build capacity. By engaging users routinely in a context that expresses a research-based approach to design, enactment and engagement, users build capacity with both the virtual and actual processes and practice that secure quality learning and teaching. They make research-based collaborative problem-solving possible. (Bain and Drengenberg 2016, p. 124)

7.5 Technology Inscribing the Professional Context

An important additional idea in explaining the technology of the self-organizing university pertains to the way an understanding of the learning and teaching context informs the design of the edge technologies employed in an SOU approach. Latour (1994) describes the way the designers of technologies employ scripts or protocols that mediate the interaction of technology and users. Like a movie or play, technologies possess a script that can prescribe the actions of those using them (Verbeek 2006). Scripts anticipate the way a technology will be used and the nature of user interaction. This concept has particular relevance to the role of technology in the SOU and specifically its agency in enabling a relationship between the professional context and the form or design of technological tools for enacting learning and teaching. In the SOU, the scripts or protocols that emerge from a university's learning and teaching model and the processes described in preceding chapters become the design blueprint for the whole-of-organization technologies employed for the work of design, enactment, and engagement with learning and teaching. Scripts inscribe the broad learning and teaching context in technologies employed in the day-to-day normal work of the university.

The example of edge technologies that follow describes this process in action. We have selected just one professional practice as a focus in order to amplify key ideas although the overall system described later in the chapter is inclusive of a whole-of-organization design. The example is unpacked to explain the role of technology in the SOU approach including the edge technology principles.

Example 7.2 Designing Cooperative Learning

Jane is developing a cooperative learning activity for her graduate course *Designing and Implementing Program Evaluations* first introduced in Chap. 4. The course serves a number of graduate programs in education and business as well as a new master of public health degree that employ program evaluation approaches. Jane is a member of a team in the school of community health in the science faculty at Grandbridge SOU that works on the course and provides feedback on the design work. She has worked at Grandbridge since it began a whole-of-organization change process using the SOU approach. Jane has some experience with the Grandbridge SOU learning and teaching model and its professional practice including the use of cooperative learning. She regularly designs courses using CL and shares feedback with peers as part of her role as a team member.

Jane is also an expert in program evaluation. In this example, Jane is developing the opening module for the course entitled *The Object of Evaluation.* Here students build a clear understanding of what is to be evaluated and the scope of the evaluation task as part of an evaluation plan that is submitted as the major assessment for the course. In Jane's module, students work in teams to become expert in four leadership tasks that are central to defining and describing what is to be the target or object of a program evaluation. Each member of the cooperative learning team has a clearly defined leadership role, and responsibility to become an expert or leader in one of the tasks. They are:

- Defining and Describing the Program
- Defining and Describing Stakeholders and Audiences
- Building a Logic Model
- Clarifying the Scope of the Evaluation

In her design, Jane reconciles the content of her module with Grandbridge SOU's protocol or script for research-based CL that includes the following achievement-related elements:

- The overall task for the module is clearly defined and broken into related component tasks;
- Each member of the cooperative learning team has a clearly defined leadership role and associated responsibility to become an expert or leader in one of the tasks;
- The deliverable from each team member necessary to complete the overall cooperative learning task is defined and described;
- The responsibility and process for sharing expertise and resources with other team members is clearly defined;
- The overall task for the purposes of assessment is completed independently.

To build her module, Jane is using a cooperative learning designer that is part of a suite of tools employed by her university to design programs, courses and learning experiences. The designer described in Fig. 7.1. captures the aforementioned achievement-related elements of the cooperative learning script in layouts and design guidance that translates the protocol into a CL learning experience.

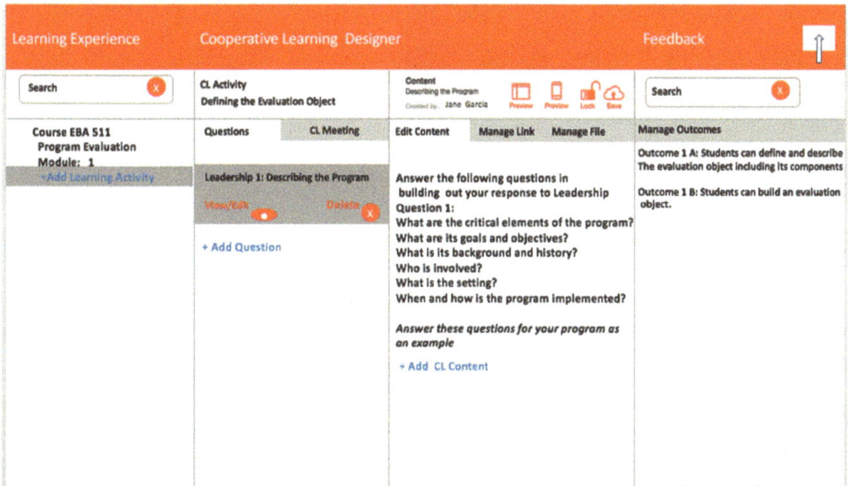

Fig. 7.1 Cooperative learning CL designer. (Republished with permission of Routledge, from Transforming the Measurement of Learning and Teaching in Higher Education, Bain and Drengenberg 2016; permission conveyed through Copyright Clearance Center, Inc.)

Figure 7.1 describes a layout from the CL designer that shows the way the protocol or script is represented in the software application. In this layout, Jane adds a description of the tasks to be assigned to each member of the cooperative learning group (leadership task). In this instance and as described in Fig. 7.1 she is entering guidance to students about the first task—describing the program. She identifies and describes the parts of a program description and notes that the student assigned to this task will build an example. Jane's work in this instance relates to the first and second bullet points in the protocol that indicate the task needs to be clearly defined while each member of the group also has a clearly defined role/task to complete that is then contributed to the group. This layout of the software helps Jane to build those leadership tasks.

As was the case with Jane's assessment task described in Chap. 4, she will ask for and receive feedback about her design as she is developing it. She may actively involve colleagues in the design process given they all have access to the tools and process.

We can unpack Example 7.1 to highlight the role of protocols, contextual interaction and a number of the edge principles in Jane's use of the CL designer.

First, Jane knows about the research-based approach to cooperative learning because it is part of the Grandbridge SOU model. The model sets the context for her engagement with CL, as a known Grandbridge SOU professional practice. The fields and design features of the tool Jane is using were developed to ensure that her design includes mutual interdependence, individual accountability, and task structure (e.g., Slavin 1996) and reflects the elements in the protocol or script for cooperative learning. While the software assists in the enactment of the cooperative learning protocol, it makes no claims and provides no restriction on the myriad ways Jane can reconcile CL practice and content and the creativity employed in designing her learning experience. Her efforts stand at the confluence of the Grandbridge SOU learning and teaching context, her professional agency, and the affordances of the CL designer. When engaged in the day-to-day work of learning design, her focus is on the module of work she is developing. The context, the designer, and her agency are represented interactively in her learning experience. She does not view the designer as a thing or a tool for automating practice.

Second, Jane's work focuses on activity that is proximal to achievement-related aspects of learning and teaching. When Jane is using the CL designer, she is setting in place the conditions for a student experience that possesses the research-based features known to improve learning outcomes. She also recognizes that the interdependence, task structure and individual accountability that are features of her CL design will produce logons, downloads, and extensive student activity in the system, although these things are of distal concern; they are at best artifacts of Jane's learning design. Jane fully understands that engagement with the LMS is driven by an interaction of her expertise with program evaluation and the research-based features of CL. Her focus is on the contextual interaction and what it produces in terms of student learning.

Jane need not have read the extensive literature or be an expert on CL to use the designer. As she interacts with the software she builds capacity with the teaching practice and the interaction between content and pedagogy (Shulman 1986) represented in her design work. Jane could build her learning experience using the pre-contextual LMS chat room to facilitate students working together in some generic sense. However, the chat room design does not possess a proximal connection to those features of CL that produce enhanced student learning outcomes and as a result will not build her capacity with the approach or realize the potential learning benefits to students.

Third, the CL designer expresses a protocol for research-based cooperative learning. It frames the way CL is perceived, interpreted, and enacted (Verbeek 2006). The CL designer possesses virtual agency as the script or protocols for CL embedded within the technology translates the Grandbridge SOU schema and learning context into the day-to-day normal work of course design across the university. In this way using the designer builds Jane's capacity and those of her peers.

Fourth, just as Grandbridge's pre-contextual effort chain model was inscribed by its LMS, the CL designer inscribes Grandbridge SOU's model and learning and teaching context. The cooperative learning designer mirrors the architecture and social construction of what Grandbridge SOU believes and does about learning and teaching. The normal gap between mission, policy, and practice is eliminated as a professional practice articulated in policy, supported and mediated by an organizational design, is instantiated in day-to-day normal work—context, professional agency, and technology interact building and extending Jane's capability and that of the organization. The CL designer also frames the foci and gaze for giving and sharing feedback as first described in Chap. 3. The feedback questions and tools shape the feedback request and the focus of feedback responses.

7.6 SOU Technologies

The example described in this chapter comes from a suite of three technologies in development for higher education learning and teaching (Bain 2012) and described in (Bain and Drengenberg 2016). They include:

Technologies for Program Design—The Programspace. The Programspace includes modules for program design including program conceptualization, standards mapping, program outcomes, assessment task, course, and module design. The modules for program design provide a starting point for the learning and teaching process that ultimately generates clear line of sight from design all the way to student performance. Analytic data reflect the connections.

Technologies for Course Design and Enactment—The CourseSpace: The CourseSpace includes modules for designing and enacting course learning activities including the pedagogical approaches and the tools to assist users deliver those approaches. The CL designer examples and feedback came from the Coursespace.

Student Technologies—The Learningspace: The Learningspace is where students engage with the learning experiences developed in Coursespaces using computer, tablet, or smartphone technology.

> All three "spaces" are connected meaning that a student can look at all aspects of the course and program design as they engage with a learning activity, making connections with course outcomes, assessment tasks and the way the activity fits into the overall program. A faculty member teaching a course can look forward to the student tools and back to the program and course spaces to situate each class session within the context of the way the course and program are designed. Each toolkit generates emergent feedback that is integrated into an analytics framework for design enactment and engagement at program, course and learning activity levels. The data for any activity, course or program can scale up to reflect data about learning activities, courses and programs at school, faculty or University level. (Bain and Drengenberg 2016, p. 131)

7.7 Implications

According to Verbeek (2006) technology has meaning within its context of use. Verbeek views the agency of technology as an interaction among context, technology, and user where the technology influences human behavior and capability interactively. In this integration, technologies become an extension of human capability and possess intentions that amplify certain functions and dampen others. They shape experience and contribute to moral decision-making (Verbeek 2006).

> Medical imaging technologies, such as MRI and ultrasound, are good examples of this. Such technologies make visible parts of the human body, or of a living fetus in the womb, that cannot be seen without them. But the specific way in which these technologies represent what they "see" helps to shape how the body or a fetus is perceived and interpreted and what decisions are made. In this way, technologies fundamentally shape people's experience of disease, pregnancy, or their unborn children. Ultrasound, for instance, makes it quite easy to determine the thickness of the nape of the neck of a fetus, which gives an indication of the risk that the unborn child will suffer from Down's syndrome. The very fact of having an ultrasound scan made, therefore, lets the fetus be present in terms of health and disease and in terms of our ability to prevent children with this disease[2] from being born. (Verbeek 2006, p. 8)

The ultrasound example illustrates ways of thinking about technology that are pertinent to education and more specifically to the self-organizing university. First, the ultrasound functions to construct a representation of the fetus showing particular developmental characteristics. That representation is shaped by the technology whereby it influences the decisions derived from its use. However, there is a much more fundamental issue at play when considering the applicability of Verbeek's example to learning and teaching in higher education.

The ultrasound imaging technology exists within a professional context. In Verbeek's example that context is the professional practice of medicine and specifically prenatal medical research and practice. The detection of a particular prenatal characteristic or problem is only possible because of the preexisting knowledge expressed by the doctor that lends meaning to what is shown on the ultrasound image. That understanding is derived from preexisting professionally controlled and evidence-based medical research and practice that also defines the skillset of the medical professional viewing the image. Meaning is derived from the ultrasound because of the preexistent understanding of what constitutes differences in normal and abnormal fetal development. The technology provides a window into that development. The interaction of the agency of the physician, the broader context of professionally controlled practice, and the affordances of the technology all contribute to the process.

We can view Jane's use of the Cooperative Learning Designer in Example 7.2 in a similar way. Jane's design for her evaluation course is an expression of her agency and role as an academic at Grandbridge SOU; the context of professional practice that exists at the university, and a technology designed specifically to articulate or express features of that context; in this instance the practice of cooperative learning. By way of comparison, the pre-contextual approach at Grandbridge described in the LMS Example 7.1 engenders a separation of agent, context, and technology because there is no-overarching professional understanding of learning and teaching to bring context, agency, and technology together interactively in professionally controlled ways. Pre-contextual Grandbridge cannot employ knowledge of cooperative learning or other research-based practices to the

---

[2]While this quote is important in its conceptualization of technology and for the purposes of this chapter we wish to note that we disagree with the representation of Down syndrome as a disease.

chat room functionality of its LMS and as a result is left hoping for an efficacious outcome derived from simply having and using chat room technology.

We can imagine the extent to which the utility of the ultrasound would be attenuated if the doctors were restricted to just having ultrasound imaging technology at their disposal without the associated professional knowledge that informs interpretation; or, how alien the cooperative learning design tool (described in Fig. 7.1) would be at Pre-contextual Grandbridge. The doctors could deploy the ultrasound to detect obvious physical characteristics and anomalies seen on the image although this is a far cry from the full scope of interpretations available to them when in possession of the full body of research and practice that comprise their professional context. While only those faculty members at Pre-contextual Grandbridge who somehow have a background experience with cooperative learning could find utility in the design tool.

We contend there is a strong similarity in using the features and functionalities of an LMS without knowledge of professionally controlled visible and comparable professional practice and the hypothetical use of the ultrasound without knowledge of prenatal development or the CL designer without knowledge of cooperative learning. Both scenarios would result in a highly attenuated role for the technology that is more likely to engender a perspective that separates context, agency and technology resulting in the focus on technology as a tool or thing.

A second issue relates to what transpires technologically in a university when there is no professional context. What fills the void or vacuum created by the lack of professionally controlled practice at Pre-contextual Grandbridge? One common response is to hyper focus on where learning happens (e.g., the chat room, classroom) or mode (i.e., online, blended, face to face) and associated technologies. We also saw this phenomenon play out in Example 7.1, where the chat room feature and the use of the LMS to expand online offerings were viewed as vehicles for delivering student success despite a lack of supporting evidence and without any sense of a professional context.

The idea here is not to suggest that mode, location, and learning technologies are unimportant. They exert a profound influence over infrastructure and staffing considerations and capital costs in higher education among other things. However, the expenditures made in these areas at Pre-contextual Grandbridge and all universities are predicated on the assumption of a distinctive and substantive effect on student learning and achievement and teaching quality—this is what the mission statements and strategic plans attest to. As we have shown throughout the book, these assumptions can only be secured when they reflect the challenging day-to-day work of a professionally controlled organization. Because the professional context for learning and teaching at Pre-contextual Grandbridge is essentially a black-box, learning and teaching quality cannot be assumed or assured. At Pre-contextual Grandbridge, the focus is fixed on the “type” of box and its “location” as surrogates for the more difficult conversation about what the box contains. Bricks and mortar or virtual become proxies for a weak understanding of the true professional learning context.

Third, when a technology like medical imaging or the tool described in Example 7.2 is employed within a professionally controlled context, it interacts in powerful ways with the physician's/academics extant skills and professional knowledge. The ultrasound enables the physician to extract a vast array of information about the professional context in which he/she operates that in turn instantiates the physician's growth and professional capability. Practice interpreting the scans builds the professional's experience and capability associated with the application of professional knowledge. Similarly, Jane can bring her extensive knowledge of program evaluation to a powerful pedagogical practice and leverage the interaction between content and pedagogy for the benefit of her students. Technology leverages expertise in the form of professional agency that in turn is amplified by the technology.

Fourth, when an expression of the learning context, technologies mediate action in ways that make it difficult and largely unproductive to sort out or separate the role of the technology from the actor and the professional context (Latour, 1994). In this contextual interaction, attention is removed from the univariate and discrete contribution of context, actor and technology to a more powerful focus on the interaction among all three. The physician does not focus on the characteristics of the ultrasound machine as a thing (Verbeek 2006) or a tool but on the image and what it shows and means. Jane does not focus on the CL designer. She is focused on building a quality learning experience. The more the technology reflects and expresses an understanding of the professional context, the less likely it is to be of singular focus or attention.

In the SOU, this interaction of context, agents, and technology is a dynamic organizational phenomenon fueled by feedback shared across the professional network of teams. This emergent feedback as we saw in Chap. 4 drives the university's capacity to act at scale. Emergent feedback is made possible by edge technology that is part of a self-organizing university's embedded design. In Example 7.2, Jane is working with the proximally important features of CL that have been shown to influence student learning. As we saw in Chap. 4, the feedback she receives about her design will focus on the same achievement-related features described in the CL script or protocol. When that feedback data becomes reportage for faculty and university use, those reports will also focus on and reflect the same proximal achievement-related factors associated with cooperative learning or any other practice in the Grandbridge SOU model.

Fifth, technologies also possess virtual agency in the way they are constructed and employed (Levy 2001). They are designed in ways that shape the gaze and action of the user. Verbeek (2006) alludes to this in the ultrasound example described earlier. The ultrasound technology focuses the gaze of the user and can differentially represent what is seen, perceived, interpreted and the decisions that follow from the interaction (Verbeek 2006). The tool Jane is using shapes a conceptualization and understanding of cooperative learning.

The form and functionality of Pre-contextual Grandbridge's new learning management system also illustrates this point unfortunately in a more counterproductive way. For example, the LMS includes breakout rooms,[3] a key feature of the new version Pre-contextual Grandbridge will adopt. The existence of this feature will shape the behavior of course developers and instructors at Pre-contextual Grandbridge. Advocates of the system will promote using breakout rooms as a best practice. Users of the system will include breakout room activity in their course designs. Students will be directed to breakout room activity and may be graded on their participation in breakout rooms. Those faculty members who do not use breakout rooms will be seen to be underutilizing the LMS and engaging in less effective practice, etc.

We also see this kind of virtual agency associated with the learning analytic functionality of the LMS. The data the system gathers becomes the data to which Pre-contextual Grandbridge pays attention. Those data are used to judge the modules developed by the relief fellows and their mentees. They will become the basis for the broader reportage of Pre-contextual Grandbridge's learning and teaching performance at all levels. Use of the LMS will be represented in key performance indicators for the university and in faculty performance management. The system focuses the gaze of the user on the analytic data deemed to be important. The design of the LMS frames the action of its users. What remains unaddressed in this cycle is the fact that there can be no expectation that the approach and its connections exert an influence on student learning.

This phenomenon also manifests at scale in the conferences conducted or sponsored by LMS developers. At these events, academics showcase exemplary use of LMS functionality and in doing so create models for other users that instantiate the value of the LMS features. Whether in course design, analytics, or in the conference examples, the functionality of the LMS instantiates a schema for what learning and teaching means by amplifying certain features, functionality, and data over other possible alternatives. In doing so, it exerts a virtual influence beyond that which occurs when using the system. "As these systems become more incorporated into everyday academic practices, they will work to shape and even define teachers' imaginations, expectations and behaviours." (Coates et al. 2005, p. 27)

As we have echoed in this chapter and throughout the book, the problem for Pre-contextual Grandbridge and other universities is the lack of any body of research indicating a student learning effect from breakout rooms or the distal analytic data derived from learning management systems. While these systems could leverage the use of evidence-based cooperative learning and more proximal analytic data, in the same way as Verbeek's ultrasound example illustrates the interaction of research and practice in prenatal development, or Jane's learning design, Pre-contextual Grandbridge's failure to articulate its learning and teaching context and build a model of evidence-based practice means that this is not possible. Its best practice will be vested solely in using the breakout room, distal analytics and other features of its new LMS. Quality ultimately becomes synonymous with the use of a system that has

[3]A virtual location for group work common to learning management systems.

little known effect on learning and teaching. Instead of measuring what it needs Pre-contextual Grandbridge measures what it has.

It is quite likely that Grandbridge University, in its pre-contextual state will not even perceive as an issue the disconnect between its strategic aspirations for learning and teaching and the capacity of its new LMS to deliver on those goals. There is a symmetry in the pre-contextual circumstances at Grandbridge and the pre-contextual form and function of the technology it chooses to deploy. In fact, if the LMS developers had combined the breakout room with CL, they would with all likelihood have created a bigger immediate problem for themselves and Pre-contextual Grandbridge. A cooperative learning breakout room design tool would require a whole-of-organization knowledge of cooperative learning in order to use the feature. This knowledge does not exist in Grandbridge's pre-contextual state. Few users in the community would know about research-based CL and how to use it in course design and implementation. Under these circumstances, a new LMS including whole-of-organization CL functionality would be viewed by the community as alien, unsuitable for normal use, difficult, excessively time consuming, and inappropriate for and insensitive to the Pre-contextual Grandbridge context and culture, etc.

By way of contrast, and illustrated in Example 7.2, the symmetry in a professionally controlled organization is one where technology, agency, and context interact in powerful and dynamic ways to improve practice and professional knowledge and release new learning and teaching potentials arising from the interaction.

Takeaways

In this chapter, we describe a different way of thinking about technology and its role in the SOU. The key takeaway propositions from this chapter are:

- There is not a strong body of evidence to support the transformational effects of technology on student outcomes in higher education and education more generally.
- Technology use in higher education follows the pre-contextual circumstances of universities and in doing so drives a separation between the agency of the technology and the agency of the user.
- To be effective, technology requires a professional context that drives the development of the scripts or protocols embedded in the design of ICT tools.
- The agency of technology resides in the interaction among context, technology, and user where the technology influences human behavior and capability interactively (Verbeek 2006).
- Edge technologies instantiate the professional context by reflecting the embedded design of an SOU in tools used for normal work.
- In the SOU edge technologies make emergent feedback possible.

References

Bain, A. (2012). Smart Tools (Versions1.0 and 2.0) Computer Software. Bathurst, NSW: Charles Sturt University.

Bain, A., & Drengenberg, N. (2016). *Transforming the measurement of learning and teaching in higher education*. New York: Routledge.

Bain, A., & Weston, M. E. (2011). *The learning edge: What technology can do to educate all children*. New York: Teachers' College Press.

Barabási, A. (2002). *Linked: The new science of networks*. New York: Perseus.

Barbour, M., & Reeves, T. (2009). The reality of virtual schools: A review of the literature. *Computers & Education, 52*(2), 402–416.

Bowker, G., & Star, S. (2000). *Classification and its consequences*. Cambridge, Massachusetts: The MIT Press.

Christensen, C., Horn, M., & Johnson, C. (2008). *Disrupting class: How disruptive innovation will change the way the world learns*. New York: McGraw Hill.

Coates, H., James, R., & Baldwin, G. (2005). A critical examination of the effects of learning management systems on university teaching and learning. *Tertiary Education and Management, 11,* 19–36.

Conway, M. E. (1968). How do committees invent. *Datamation, 14*(5), 28–31.

Cuban, L. (2003). *Oversold and underused: Computers in the classroom*. Cambridge, MA: Harvard University Press.

Drengenberg, N., & Bain, A. (2016). If all you have is a hammer, everything begins to look like a nail—How wicked is the problem of measuring productivity in higher education? *Higher Education Research & Development*. doi:10.1080/07294360.2016.1208640.

Dynarski, M., Agodini, R., Heaviside, S., Novak, T., Carey, N., Campuzano, L., et al. (2007). Effectiveness of reading and mathematics software products: Findings from the first student cohort. (Publication No. 2007–4005). Retrieved April 27, 2007, from Institute of Education Sciences, U.S. Department of Education, Web site: http://ies.ed.gov/ncee/pdf/20074005.pdf

Facer, K., & Sandford, R. (2009). The next 25 years? Future scenarios and future directions for education and technology. *Journal of Computer Assisted learning, 26,* 74–93.

Guri-Rosenbilt, S., & Gros, B. (2011). E-Learning, confusing terminology, research gaps and inherent challenges. *International Journal of E-Learning and Distance Education, 25*(1). Retrieved from http://www.ijede.ca/index.php/jde/article/view/729/1206.

Hattie, J. (2008). *Visible learning: A synthesis of over 800 meta-analyses relating to achievement*. New York: Routledge.

Kimmons, R. (2015). Online system adoptions and K-12 academic outcomes. *Journal of Computer Assisted learning, 31*(4), 378–391.

Latour, B. (1994). On technical mediation: Philosophy, sociology, genealogy. *Common Knowledge, 3*(2), 29–64.

Levy, P. (2001). *Cyberculture*. Minneapolis: University of Minnesota Press.

Means, B., Bakia, M., & Murphy, R. (2014). *Learning online: What research tells us about whether, when and how*. New York: Routledge.

Rao, A. (2012, January 3). 10 Educational Technology Quotes. Retrieved from http://teachbytes.com/2012/03/01/10-educational-technology-quotes/.

Santoro, L., & Bishop, M. (2010). Selecting software with caution: an empirical evaluation of popular beginning reading software for children with early literacy difficulties. *Computers in the Schools, 27*(2), 99–120.

Shapley, K., Sheehan, D., Sturges, K., Caranikas-Walker, F., Huntsberger, B., & Maloney, C. (2009). *Evaluation of the Texas Technology Immersion Pilot: Final outcomes for a four-year study (2004–05 to 2007–08)*. Austin, TX: Texas Center for Educational Research.

Shulman, L. S. (1986). Those who understand: Knowledge growth in teaching. *Educational Researcher, 15,* 4–14.

Silvernail, D. (2007). The impact of the Maine Learning Technology Initiative on teachers, students, and learning. Retrieved from http://usm.maine.edu/cepare/mlti.htm.

Slavin, R. E. (1996). Research on cooperative learning and achievement: What we know, what we need to know. *Contemporary Educational Psychology, 21,* 43–69.

Smith, R., Clark, T., & Blomeyer, R. (2005). *A synthesis of new research on K–12 online learning*. Naperville, IL: Learning Point Associates.

Steenbergen-Hu, S., & Cooper, H. (2014). A meta-analysis of the effectiveness of intelligent tutoring systems on college students' academic learning. *Journal of Educational Psychology, 106*(2), 331–347. doi:10.1037/a0034752.

Trucano, M. (2005). *Knowledge maps: ICTs in education.* Washington D.C.: InfoDev, The Information for Development Program. Retrieved from http://www.infodev.org/en/Publication.8.html.

Verbeek, P. P. (2006). 'Materializing Morality—Design ethics and technological mediation. Retrieved from https://www.utwente.nl/bms/wijsb/organization/verbeek/materializing-morality.pdf.

Webopedia. (2016). Web 2.0. Retrieved from http://www.webopedia.com/TERM/W/Web_2_point_0.html.

Weston, M., & Bain, A. (2014). Bridging the research-to-practice gap in education: A software-mediated approach for improving classroom instruction. *British Journal of Educational Technology*. doi:10.1111/bjet.12157.

Weston, M., & Bain, A. (2010). The end of techno-critique: The naked truth about 1:1 laptop initiatives. *Journal of Technology, Learning and Assessment, 9*(6), 5–26.

Windschitl, M., & Sahl, K. (2002). Tracing Teachers' use of technology in a laptop computer school. *American Educational Research Journal, 39*(1), 165–205.

Youssef, A., & Dahmani, M. (2010). The impact of ICT on student performance in higher education: Direct effects, indirect effects and organisational change. Paper Presented at The First International Conference on e-Learning for All, Hammamet, Tunisia. Retrieved from http://pf-mh.uvt.rnu.tn/45/.

Chapter 8
Learning and Teaching Productivity

8.1 Introduction and Overview

In this chapter, we take up learning and teaching productivity in the Self-Organizing University (SOU). We build upon the work of preceding chapters to show how the SOU demystifies quality and efficacy as it relates to productivity in university learning and teaching. To open the chapter, we make the case for improved productivity. We then employ an example from Pre-contextual Grandbridge to illustrate the challenges associated with measuring productivity using an effort-chain approach. We show how productivity determinations can be legitimized in the SOU when the learning and teaching context is understood. Our second example illustrates this approach in practice at Grandbridge SOU. We conclude the chapter with a discussion of the implications of the SOU approach for productivity measurement in higher education.

8.2 Pressure

Between 2003–04 and 2013–14, prices for undergraduate tuition, fees, room, and board at public institutions in the United States (US) rose 34%, while prices at private nonprofit institutions rose 25%, after adjustment for inflation (USDOE 2014). Costs are expected to continue to grow at 5% per year yielding an estimated figure of 40,935 dollars for a 4-year degree from a public institution and 90,576 dollars for the same degree from a private institution by 2030 (Badkar 2014). As a corollary, the power of a degree in returning a guarantee of work and income commensurate with or relative to cost has become much less clear (Barrett 2015).

> The salary gap between those with only a high-school diploma and those with a university degree is created by the plummeting value of the diploma, rather than by soaring graduate salaries. After adjusting for inflation, graduates earned no more in 2007 than they did in

A. Bain and L. Zundans-Fraser, *The Self-organizing University*,
DOI 10.1007/978-981-10-4917-0_8

> 1979. Young graduates facing a decline in earnings over the past decade (16% for women, 19% for men), and a lot more debt, are unlikely to feel particularly cheered by the argument that, over a lifetime, they would be even worse off without a degree than with one. (The Economist 2012, para, 13)

Barrett (2015) describes a pending tipping point where the rapid escalation in tuition costs ultimately fail to offset the prospective earning gains making college no longer worth the price of admission. Cost concerns are exacerbated by the fact that many students are not actually completing the degrees for which they enroll. Only 35% complete their degrees in five years at public universities (USDOE 2014). Concerns about cost, buying power, and completion have increased pressure on universities to show both the efficacy, efficiency, and overall productivity of what they are providing to students and to demonstrate improvements in those areas. As illustrated throughout the book, we contend that the core issue associated understanding with escalating cost and poor completion is the inability to understand what learning and teaching means at scale.

8.3 Pre-contextual Productivity

In economic terms, productivity is a measure of the efficiency of a person, machine, factory, system, etc., in converting inputs into useful outputs. Productivity is computed by dividing average output per period by the total costs incurred or resources (capital, energy, material, personnel) consumed in that period (Businessdictionary 2016).

The computation of productivity has been applied to higher education in the form of single and multi-factor productivity ratios (Massy 2016). Single factor productivity models involve a single input and output measure; for example, relating student credit hours to staff full-time teaching load equivalency (FTE). Single factor models are attractive in their simplicity although they often overlook the full costs of teaching including the cost of resources, differences in faculty experience/qualification, and the cost of technology (Massy 2016). They also mask faculty overload where increases in teaching loads due to cost cutting may be reducing the quality of teaching despite the appearance of improved productivity (Massy 2016).

Multi-factor models broaden the scope of inputs and outputs. An example of a dynamic multifactor model is described in the *National Research Council Report, Improving Measurement of Productivity in Higher Education (NRCR)* (Sullivan et al. 2012) introduced in Chap. 1. The NRC propose a productivity measure entitled the Törnqvist index which is described as:

$$\text{Productivity index } [\Delta t] = \text{Output index } [\Delta t] \div \text{ Input index } [\Delta t]. \text{ (p. 7)}$$

The Törnqvist index defines productivity as the change in outputs obtainable from the input changes observed over Δt. The input and output indices represent changes in the physical quantities over the time measured. Productivity change, in turn is calculated from the ratio of successive productivity indices:

$$\text{Productivity change } [\Delta t1 \text{ to } \Delta t2] = \text{Productivity index } [\Delta t2] \div \text{Productivity index } [\Delta t1] - 1.$$

In the higher education context, the outputs in the model refer to credit hours adjusted for learning that takes place on a course-by-course basis, program completions, and a "sheepskin effect" to take into account the value added via the completion of a degree, beyond the completion of individual units. The inputs are determined based on the cost of labor (faculty teaching), the cost of materials and the use of physical capital (e.g., land, buildings and equipment). The base model is augmented by differentiating labor categories in various ways (e.g., distinguishing between full-time and part-time faculty) and the type and cost of degree (e.g., graduate/undergraduate and cost of delivering a specific type of degree e.g., medicine/arts) (Sullivan et al. 2012). The Törnqvist Index expands upon single factor models to include a more drilled down account of the contributions to productivity while also reflecting change over time that can be used for planning and evaluation purposes.

Activity-Based Costing (ABC) is an elaboration on multi-factor indices and involves an even closer examination of teaching activities (e.g., student projects, class sessions, out of class communications), credit hours, resources (e.g., facility use, technologies), and unit costs (e.g., faculty time). The ABC approach includes a more fulsome account of what goes on in learning and teaching at universities. Massy (2016) describes course-based ABC as the approach that "really works" (p. 120). It leads to a clearer understanding of what a module or course of instruction includes while its unitized approach makes possible the examination of a university's teaching portfolio at scale (i.e., learning activities, module, course, portfolio of courses). ABC represents a clear improvement on single and existing multifactor models as it begins to articulate in more detailed categories what constitutes the learning and teaching inputs and outputs in a productivity ratio (Massy 2016).

> It (ABC) satisfies the criteria for analysing what economists call the teaching production function by getting to the heart of the relationship between resource utilization and output. (Massy 2016, p. 123)

From the perspective of the pre-contextual university the aforementioned description of ABC is a complete and workable solution. ABC is perfectly symmetrical for a set of circumstances whereby academics, economists, and university leaders are unaware of, unconcerned by, or unconvinced about the professional practice of learning and teaching at scale; or have simply placed it in the too hard basket when addressing productivity. Under these circumstances, establishing how much time is spent preparing, marking, or giving lectures is a fully adequate approach to determine productivity inputs while drilling further into these aspects of teaching production generates even better data. In Example 8.1 we show how the productivity model plays out at Pre-contextual Grandbridge University.

Example 8.1 Pre-contextual Productivity at Grandbridge

President Harper is keen to quantify learning and teaching at Pre-contextual Grandbridge. He wants to demonstrate in dollar and cents benefit, the expected productivity gains associated with the university's reorganization described in Example 5.1. He is also concerned about the high costs associated with the new assessment model (Example 6.1) and its failure to yield returns on the university's substantial investment. An additional overarching concern is Pre-contextual Grandbridge's inability to improve its on time graduation rate of 19% for a four-year diploma. He believes better data would make Pre-contextual Grandbridge more accountable and help resolve these problems.

Pre-contextual Grandbridge's Chief Financial Officer (CFO) wants to introduce a new software program that will help the university understand its costs and inform planning decisions. Research undertaken by the CFO has identified a company offering a customized and integrated software and consulting package for determining higher education learning and teaching productivity. The approach is based on a multi-factor productivity model that employs credit hours and graduation as outputs. The CFO is especially enthusiastic about the way the approach employs historical fiscal performance to create a predictive model. The company has already built a list of high profile clients and is developing a consortium of users. A forthcoming national conference for members is in the planning stages. Both President Harper and his CFO believe the approach will show positive results for the strategic planning initiatives at Pre-contextual Grandbridge. While the software application and consulting services represent a significant expense, President Harper approves the purchase.

After twelve months of use, the reaction from Pre-contextual Grandbridge's CFO to the new software is one of great enthusiasm. The software is well matched with the administrative re-organization described in Example 5.1. The re-organization has eliminated the idiosyncratic faculty support structure replacing it with a whole of university approach. The input costs are now stable and consistent across faculties, and are well-suited to the software's input parameters. Since its introduction, the software has brought a new level of clarity to the analysis of the university's fiscal performance. Data that existed in separate spreadsheets from different offices can now be displayed interactively and modelled dynamically. The CFO can examine the labor costs associated with faculty (full and part-time) and the income they generate, the differences between online and face-to-face tuition costs, current and projected enrolment, completion rates, and productivity by school, discipline, campus, faculty, and the university overall. This all happens using one platform and a dashboard model for easy comparison and interpretation of inputs and outputs.

For the CFO, the data is symmetrical for Pre-contextual Grandbridge's understanding of its learning and teaching context. The software paints a picture for the university that is clear and complete. He can employ the program's dashboards to see exactly how much a Pre-contextual Grandbridge unit of instruction and a full degree costs per student; the variability across faculties in inputs and outputs; the productivity of individual faculty members, schools, and faculties and the university as a whole; the variability in resource consumption and facility use, and differences in the cost of online versus face-to-face courses and degrees. Most important, the model and software make possible the projection of the fiscal health of the university moving forward based on costs, enrolment as well as historical, current, and projected credit completion and graduation rates. The approach is providing a much clearer picture of the cost structure at Pre-contextual Grandbridge and the relationship between inputs and outputs. The software also shows the new re-organization seems to be producing some cost savings despite there being no demonstrable change in graduation rates from faculty to faculty.

The CFO of Pre-contextual Grandbridge registers for the inaugural national conference, an event designed for those members of the consortia using the consulting company's software. He attends a session offered by one of his peer CFOs and an executive of the software and consulting company entitled "Unbundling the University.[1]" The session describes how members of the consortium with their shared productivity model and software platform can unbundle the services they offer to students. The presenters explain how a student may receive library services from one university, online services from another, and content from multiple members of the consortium. Under these circumstances, members of the consortium could become providers for each other achieving better economies of scale while focusing upon and sharing their strengths for individual institutional benefit.

The CFO is excited about the concept. He knows that Pre-contextual Grandbridge is struggling to meet its strategic planning goals of leadership in the field (Example 2.1); the expensive assessment initiative (Example 6.1) is not producing the desired results, and graduation rates remain unchanged. He also believes Pre-contextual Grandbridge's long history in distance learning can be turned to advantage in an unbundled world. The CFO thinks unbundling may provide a new way forward to maintain the university's fiscal viability.

Example 8.1 described the way productivity is configured from a pre-contextual perspective and how that played out at Grandbridge University. What follows is an account of productivity in the SOU approach and its implications for learning and teaching.

8.4 Genuine Learning and Teaching Productivity

A legitimate model of learning and teaching productivity requires the following:

1. Inputs that reflect known attributable relationships between teaching and student learning outcomes. These are the same relationships that define the learning context described in Chap. 1 and are referred to in subsequent chapters;
2. Outputs that reflect valid and reliable measures of student outcomes against the standards and professional requirements of a degree program, course, or learning experience;
3. A known empirically demonstrable relationship between inputs and outputs;
4. Workable distinctions (Drengenberg and Bain 2016) in the quality of learning and teaching work specifically related to the inputs and outputs;
5. Emergent feedback systems that can produce the valid and reliable measurement of 1-4.

[1]See Staton (2012). Disaggregating the components of a college degree. American Enterprise Institute. Retrieved from http://www.aei.org/wp-content/uploads/2012/08/-disaggregating-the-components-of-a-college-degree_184521175818.pdf for an explanation of unbundling or disaggregation.

We can use the five productivity requirements as a guide to unpack Example 8.1. Pre-contextual Grandbridge's CFO is excited about being able to see inputs like faculty labor, material expenses etc., related to credit hours completed and graduation using the software dashboards. The CFO is being afforded greater professional control within a context he understands and for which he is responsible. The new model and software provide a much more expansive view of costs and greater detail about what teaching involves at Pre-contextual Grandbridge. However, what is unexplained is the quality and efficacy of what those faculty members are doing with their teaching time and resources. For reasons described extensively in the preceding chapters, Pre-contextual Grandbridge's effort chain approach means it has a poor understanding of its learning and teaching context and the relationships that need to exist in order to establish the quality and efficacy of its inputs to learning and teaching.

This issue is not within the CFOs sphere of influence or responsibility. He simply assumes the data produced by the software reflects "good learning and teaching" as expressed in the Pre-contextual Grandbridge strategic plan and mission statement. If this is not the case, then the Vice-President (Academic), the Human Resource Division or someone else needs to address the issue. The CFO's assumptions reflect the way responsibility is partitioned in an organization that does not understand what learning and teaching means in a professionally controlled sense. As we saw in Chap. 5, Pre-contextual Grandbridge deals with its weak understanding by defining categories of responsibility. It expects to build collective capacity by assigning responsibility to different roles and locations on the organizational chart. However, the result of this approach is more likely to be the fragmentation of understanding and responsibility leading to the CFO's assumption that the university's productivity data represent something that works; if this is not the case, then the problem lies elsewhere and is someone else's responsibility to fix.

Second, in relation to outputs, serious questions exist about whether credits completed and graduations are legitimate measures of student outcomes. We know that students can progress at college without substantive development in their learning. Graduation and/or credit accumulation and the attainment of learning outcomes are not synonymous. For example, a national US survey showed that the literacy of college-educated citizens declined between 1992 and 2003. "Only a quarter were deemed proficient, defined as "using printed and written information to function in society, to achieve one's goals and to develop one's knowledge and potential. Almost a third of students these days do not take any courses that involve more than 40 pages of reading over an entire term" (The Economist 2012, para. 9). The extent to which the owners of degrees comprised of this type of experience are making a major contribution to a nation's productivity, well-being and success is an interesting question.

More expansive evidence in support of the tenuous relationship between inputs and outputs in the pre-contextual university was provided in Chap. 1. There we noted that gains in student learning from entry to graduation are frequently small and there is much more variability in student performance within universities than between them indicating the limited contribution of a university's unique approach to student learning (ACT 2009; Liu 2008; Pascarella and Terenzini 2005). The issues associated with the quality and reliability of what students learn and master at pre-contextual universities are exacerbated in circumstances when the degree awarded is a professional qualification. Those degrees require students to meet professional standards of practice for a given field (e.g., accountancy, teaching, medicine etc.). The use of credits completed and graduation may in fact mask learning and teaching problems that could indicate a lack of readiness for the professions.

Third, when the inputs and outputs that constitute the numerator and denominator in a productivity index are inaccurate or ambiguously defined the effect is to amplify the ambiguity in the ratio overall. Imagine running a factory where you knew when the workforce clocked in and out but had limited understanding of what they did with their time, what they produced, and how those inputs and outputs covaried. Again, if the factory was not functioning well, with such a distal understanding of inputs, outputs, and the relationship between them, the business could only speculate about whether the factory was productive. There would also be little guidance available to address its problems.

Fourth, at the heart of Pre-contextual Grandbridge's limited problem-solving capability is its inability to discern what constitutes workable distinctions in performance (Drenegenberg and Bain 2016). This means establishing in valid and reliable ways the differences in the quality and efficacy of efforts to design and deliver learning and teaching experiences. Pre-contextual Grandbridge cannot show how the time and effort expended by faculty members produces work of differential quality and effectiveness. The inability to differentiate performance is a consequence of an effort-chain approach and the reliance on distal measures like time, load, and credits as primary data sources for measuring productivity. The latter are poor proxies for workable distinctions in the quality of learning and teaching (Drengenberg and Bain 2016). Further, drilling down and partitioning teaching activity provides no more clarity about workable distinctions.

Finally, and most problematic, while the Pre-contextual Grandbridge CFO is excited about the way new software dashboards produce such a complete pre-contextual picture of productivity, there is no data or dashboard representation of the emergent feedback (Chap. 4) about the design and delivery of learning and teaching. This is the data that could discern quality, set standards, and determine workable distinctions at scale in order to improve student outcomes and lend underpinning meaning to productivity indices. While Pre-contextual Grandbridge has a more complete picture of what it is doing, it does not possess the feedback to improve the quality, efficacy and productivity of learning and teaching.

8.5 The Transposition Problem

The pressure to account for the costs of higher education has created a problem of transposition at Pre-contextual Grandbridge, where an effort is made to quantify learning and teaching productivity without establishing or understanding the relationships between inputs and outputs. What we see in the single and multi-factor pre-contextual productivity indices and Example 8.1, is the way productivity methodologies employed in fields with known professional control (Bowker and Star 2000) are simply transposed for use in the higher education context. This happens without addressing the legitimate preconditions necessary to meaningfully determine learning and teaching productivity in a higher education setting. The transposition problem is effectively illustrated by returning to the unbundling element of Example 8.1. According to Horn (2014):

> In every industry, the early successful products and services often have an interdependent architecture—meaning that they tend to be proprietary and bundled. The reason for this is that when a technology is immature, in order to make the products reliable or powerful enough so that they will gain traction, an entity has to wrap its hands around the whole system architecture so that it can wring out every ounce of performance. As a technology matures, however, it eventually overshoots the raw performance that many customers need. As a result, new disruptive innovations emerge that are more modular—or unbundled—as customers become less willing to pay for things like power and increased reliability but instead prioritize the ability to customize affordably by mixing and matching different pieces that fit together according to precise standards. (para. 2 and 3)

Horn's, description of more or less mature technologies and unbundling offers a profound insight to the transposition problem in the pre-contextual university. Unbundling presumes higher education learning and teaching to be a mature technology ready to move beyond a whole system architecture. The reality is that pre-contextual universities are yet to develop sufficient understanding of what learning and teaching means to build the foundational architecture for the kind of technology Horn describes as a pre-condition for unbundling. As shown throughout the book, a mature technology stands at the confluence of standards of practice, technology, governance, and human resource factors etc., in a self-organizing university. These are the interdependent elements required to make a university's approach to learning and teaching reliable and powerful (Horn 2014) at scale.

There is little doubt that a pre-contextual university could outsource its library or online learning services to a peer institution. However, as we have seen in the preceding chapters, doing so means little in terms of student outcomes unless those services somehow represent an understanding of the learning context and a value proposition that indicates a differential effect of those services on student learning. These questions remain unanswered in the unbundling conversation. As with so many other innovations, higher education simply skips the issues related to the maturity, efficacy and quality of its core activity, and proceeds to impersonate more mature fields; in this case the unbundling and modularization of learning and teaching services. It is clear from the case made in the preceding chapters and the extensive literature on the state of learning and teaching in higher education that the pre-contextual university is far from a position where its learning and teaching capability is "overshooting the raw performance customers need (para, 3)." At present, unbundling really means disconnecting and modularizing the services that drive the effort chain which have not been shown to exert an influence on quality, efficacy, or genuine learning and teaching productivity.

The most compelling way to illustrate the consequences of the transposition problem is to consider what the new software model described in Example 8.1 would offer up to solve or address Pre-contextual Grandbridge's productivity problem of low 35% five-year student completion rate or its assessment problem described in Chap. 6? Pre-contextual Grandbridge may be able to use the model to identify differences in costs among faculties and schools or link performance to student entry characteristics (the latter being the biggest predictor of performance) and establish progress rates in online courses. However, despite the CFOs satisfaction with what seems to be such a complete picture of Pre-contextual Grandbridge, there is little information that would indicate how to improve learning and teaching, address the low rates of graduation, or any other real learning and teaching productivity problem.

As we have seen throughout the book, it is not whether there *is* a lecture that is important but how a learning experience is designed and delivered to maximize student outcomes. The true value of preparation time *is not* about whether it happens but what transpires as a result. Imagine determining the productivity of a medical or legal practice by establishing whether patients or clients are seen, hours/appointments billed, and resources consumed, without knowing whether the use of time and consumption of resources results in the successful resolution of cases or whether patients regained their health. If a university wants to save money associated with the cost of lectures or any other learning and teaching element, it has no standards, no term of reference to make those decisions beyond looking at others doing the same thing. This absence of understanding about efficacy and quality, while acknowledged is simply accepted as the normative state of affairs in pre-contextual productivity modelling in higher education (Sullivan et al. 2012).

The transposition problem results in an immense gap between what is expected of the software as described in Example 8.1 and what it can deliver in terms of addressing Pre-contextual Grandbridge's learning and teaching productivity needs. With new productivity modeling and software, the CFO may have a better understanding of how and where money is spent at Pre-contextual Grandbridge. However, it is no surprise as indicated in Example 8.1 that Pre-contextual Grandbridge may be improving its pre-contextual learning and teaching "productivity" without improving its learning and teaching.

The example of Pre-contextual Grandbridge's efforts to take a more disciplined and sophisticated pre-contextual approach to productivity is another echo of Conway's law. As with standards, promotion frameworks, and technologies for learning and teaching, the approach to productivity is symmetrical for the effort chain and what learning and teaching means at Pre-contextual Grandbridge. There is no deep engagement with the view that improving learning and teaching productivity requires an understanding of what is being produced and the quality of the inputs required to produce it. Distal approximations of those inputs and outputs suffice and as a result produce indicators of productivity that have not much to do with the quality or efficacy of learning and teaching. There is also little doubt that Pre-contextual Grandbridge will continue to follow its data when it comes to planning and decision-making. As we saw in Chap. 2, the architecture and social construction (Conway 1968) of Pre-contextual Grandbridge, including its productivity data will shape its actions moving forward.

8.6 Productivity in the SOU

In the preceding chapters, we have made the case for improving student outcomes through professionally controlled learning and teaching practice (Bowker and Star 2000). In Example 8.2 and the discussion that follows, we propose that professional control and an understanding of the learning and teaching context can also improve the fiscal performance of a university by linking expenditure to quality and efficacy. The example will demonstrate how the best intents of improved student learning in the pre-contextual effort chain university can be made operational in ways that turn out to be better for students and better for business in the SOU.

Example 8.2 Grandbridge SOU builds a Learning and Teaching Productivity Model

The leadership team at Grandbridge SOU is eager to capture the benefits of the university's emergent schema for learning and teaching. Professional control at Grandbridge SOU is offering up some exciting new potentials for understanding the costs associated with learning and teaching and the university's productivity. Grandbridge SOU is confident that its SOU approach and specifically the emergent feedback it is accumulating makes possible a new way of thinking about productivity. As a result, the leadership team decides to build a productivity index that meets the requirements of the SOU and reflects Grandbridge SOU's capacity to link cost and educational productivity with quality and efficacy. The university produces the following SOU Productivity Index:

$$\frac{\text{Module Design} + \text{Module Delivery}^{(O/F)}}{\text{Student Module Outcomes} \times \text{Total Credithours}}$$

Key

Module Design = The total cost of building an approved module of instruction[2] in the SOU software including faculty, time, resources, professional development, technology and other infrastructure.

[2]An approved module means that the module has been subject to the emergent feedback approach described in preceding chapters and has met accepted standards for quality and efficacy.

Module Delivery = the cost of teaching the module including faculty time, resources, technology and infrastructure costs). The (O/F) designation reflects the difference in cost structure depending on whether the module is delivered online or face-to-face.

Student Module Outcomes = (average of percent outcomes mastered + percent course grade)/n

Total Credit hours = the credit hours for the course multiplied by the number of students enrolled.

The leadership team at Grandbridge SOU believe the index can drive a whole new range of benefits related to the way Grandbridge SOU expends resources. It can show in clear and unambiguous ways how much a module of learning and teaching costs, the expertise required to develop one, and whether the resources consumed can be attributed to student outcomes. The leadership at Grandbridge SOU also believes it can show definitively that its students learn more in less time after gathering five-years of emergent feedback data across courses and programs throughout the university. Working with professional bodies in five fields, and state regulators, Grandbridge SOU has successfully demonstrated that its model of learning and teaching can accomplish in three years what other universities are doing in four. As a result, it has secured agreements to qualify students in less time. The process has not been easy requiring the development of unequivocal longitudinal evidence and extensive negotiation to shift traditional attitudes in conservative organizations about time as a constant in higher education and the longstanding view that rigor is synonymous with a four-year qualification.

The marketing team at Grandbridge SOU is excited. After years of trying to put a different spin on the tired effort chain story at the former pre-contextual version of the university, the team has something new and more powerful to work with. It is about to launch a large national and international marketing campaign. Five years into the Grandbridge SOU approach, the productivity index and the university's emergent feedback software is producing the kind of data other effort chain university's wish they could produce.

Using the new index and its underpinning emergent feedback, the marketing team has taken the rhetorical claims from the marketing materials and webpages of the country's ten leading universities and produced Grandbridge SOU's versions that tell the same story in a profoundly different way. The highlights in competitors' marketing messages like "world standard" "great teachers" and "quality education" are translated into personal stories of student success. What makes these stories different is the way they are punctuated with real evidence of the way the generic aspirational claims of competitors are legitimized at Grandbridge SOU. This includes cases of students finishing their degrees faster because they learn more in less time and the implications for degree programs and the university as a whole. The team believes the marketing case along with the new approvals are so strong that it expects a marked expansion in the university's enrolments especially in online courses.

Grandbridge SOU's clear understanding of the relationship between its inputs and outputs has enabled the team to produce a powerful and accessible new story about what makes Grandbridge SOU different and more importantly better at learning and teaching. The centrepiece of the Grandbridge SOU campaign is the motto *"better, faster, affordable."* Grandbridge SOU knows the marketing campaign is an expression of a deep and fulsome capability to demonstrate learning and teaching productivity. This is the university's differentiator and the path to securing its future.

Grandbridge SOU has crossed a border. It understands how its model and approach effects learning and is positioning itself to realize the immense economic potential of that understanding. Grandbridge SOU's success is creating a disturbance in the field. After a recent presentation by President Harper at a national conference on University Performance Standards, a leading member of the national accrediting body was heard to say—"this changes everything."

We can unpack Example 8.2 to ascertain whether the Grandbridge SOU Index meets the requirements of a legitimate productivity model. The first requirement pertains to the validity of the inputs as they relate to learning and teaching outcomes. In the SOU, this happens in six ways. The software tools (Chaps. 3, 4 and 7) being used to develop the modules represented in the index incorporate practices known to exert an attributable effect on learning. The use of the tools is subject to constant feedback from peers focusing on workable distinctions in the quality of design (Chaps. 3 and 4). The qualification of the peers who provide feedback is based upon progression in a career framework that focuses on expertise with the things know to influence the quality and efficacy of module design (e.g., Figs. 3.7 and 6.1). The ongoing emergent feedback makes possible the development of standards for module design against which all modules are judged. Those standards determine whether a module is approved. The governance approach at SOU (Chap. 5) employed to approve the module applies the standards and workable distinctions in the module design. Most important, Grandbridge SOU can demonstrate the effect of those modules in their design and delivery on student outcomes.

Many of the same conditions exist for the delivery of the modules. Delivery from an SOU perspective means implementing the validated design as articulated in the tools. The delivery process, like design is also the subject of peer evaluation by individuals qualified to evaluate what they are observing (Example 4.2). As illustrated in Example 3.2, the questions asked of students in measures employed to evaluate the delivery of the module, reflect the attributable relationships among learning and teaching expected at Grandbridge SOU. The emergent feedback about delivery, like the feedback shared for design generates empirically derived standards for practice that inform determinations of quality and efficacy and workable distinctions in performance. The governance process draws upon emergent feedback about the quality of delivery as a factor in ongoing review and approval decisions. Grandbridge SOU can demonstrate the effect of module delivery on student outcomes. The module whether in design and delivery is a validated expression of the learning and teaching context at Grandbridge SOU.

The second requirement of a legitimate productivity index pertains to outputs. As we saw in Example 8.2 the denominator in the index expresses a process designed to assure the attainment of student outcomes in Grandbridge SOU courses and programs. Course and program outcomes are mapped to professional standards while performance in the attainment of those outcomes is factored into the numerator in terms of course grade. The direct inclusion of student outcomes is made possible because of the way in which the Coursepace software tools in the SOU connect each high level outcome back to a course assessment which in turn is linked to a program standard and productivity requirement. This means that the denominator in the SOU Index corresponds directly to a valid measure of student performance in terms of outcomes achieved and grades. We saw an example of this play out in Example 5.2 where the analytic capabilities of the Grandbridge SOU software made possible the connection between an assessment issue and student outcomes.

Third, the connection between inputs and outputs in the Grandbridge SOU model begins with the learning and teaching context and the way it is instantiated by the embedded design process described in Chap. 3. As we described in that chapter, embedded design builds into the normal work of the organization the connections between practices, attributable relationships, standards, and workable distinctions that drive determinations of quality and efficacy. The Grandbridge SOU productivity index is another way to express the relationship between the design and delivery of a course and student outcomes that can be mapped all the way back to the university's commitments and policy. The relationship between numerator and denominator on the SOU productivity index is a self-similar echo of every day work at Grandbridge SOU as faculty seek to design and define modules based on their performance as indicated by student outcomes. The Index as described is symmetrical for the Grandbridge SOU organizational design; is underpinned by an understanding of the learning context at Grandbridge SOU, and reflects an attributable empirical relationship between learning and teaching.

Fourth the definition of inputs and outputs not only makes a determination of quality and efficacy possible, comparable and visible variance in practice can be assessed against standards and the expected attributable relationships to determine workable distinctions in performance. The data employed to detect those workable distinctions in practice is the product of the fifth requirement, an emergent feedback system (Chap. 3) that makes standards, workable distinctions and determinations of quality and efficacy subject to constant shared feedback for dynamic change.

8.7 Implications

Grandbridge SOU has set in place the conditions to determine efficacy and quality. It can answer questions like: Has a learning experience been designed using quality evidence-based practice? Is the use of those practices translating into effective student learning? Does the module meet standards for design at Grandbridge SOU? Do all students meet the subject and program learning outcomes and is their performance indicative of the skills required to meet the professional standards of their degree? Does the feedback from students about the module reflect those standards?

Further, the fundamental productivity issues of cost and cost savings can be situated within the answers to the aforementioned questions. We can establish how much a learning and teaching module costs based upon an understanding of the learning and teaching context and can relate cost to quality and ultimately productivity.

Most important, Grandbridge SOU can answer these questions at a whole-of-university scale. The existence of known standards, attributable relationships and workable distinctions means that learning and teaching at Grandbridge SOU can be subject to the performance studies and determinations of price competitiveness applied in other fields (Sullivan et al. 2012). These are the conditions required to make the connection between quality, efficacy and productivity at scale.

The more Grandbridge SOU understands what it is doing, the more it is able to improve the inputs and outputs in its index (e.g., as faculty capacity improves it learns more effective ways to expend resources in pursuit of better course design and delivery or in refining course and program outcomes). Those influences can be expected to improve standards, refine workable distinctions and improve student outcomes influencing both numerator and denominator in the index. In doing so, Grandbridge SOU can refine its tools, its professional development, its promotion frameworks, commitments and policy dynamically as it refines the way all of its systems, methods, and tools instantiate its understanding of the learning and teaching context.

Instead of unbundling its services, Grandbridge SOU is leveraging the power derived from the totality of its design and organizational architecture. Its educational value proposition lies in the deep connections among those things other universities are seeking to unbundle in generic ways. Grandbridge SOU believes it will be positioned strongly to take a distinctively branded and genuinely different courseware product to new markets as it consolidates the strength of its existing mission and performance, while its design and especially its emergent feedback capability can maintain the differential.

Educational reform and innovation is rarely related to cost savings and efficiency. For many of the reasons described here, change or improvement is synonymous with spending more. A main idea of this chapter is that the poor understanding of the relationship between the inputs and outputs in pre-contextual productivity indices is symmetrical for the poor understanding of the relationship between learning and teaching. Both are bad for business. The predictive drumbeat of tuition increases reflects the immature state of learning and teaching productivity. While multi-factor productivity models, including the ABC approach make a concerted effort to capture more about the activity of learning and teaching none help a university to understand whether resources expended on learning and teaching work for expected effect.

No university knows at scale whether its assistant professors are better at designing and delivering learning experiences than its professors or how long it takes to build mastery of its learning and teaching processes. While lectures continue to be built and time is consumed incorporating new technologies into online learning experiences there is no evidence to link the time and resource expenditure on those lectures or online experiences to their impact on student learning. Tutors and tutorials seek to unpack big ideas and capture the perceived benefits of a smaller teacher to student ratio without ever knowing whether those expected productivity returns materialize in terms of better student learning beyond the session-by-session reportage from student satisfaction measures. Research in higher education would suggest those smaller classes are only modestly correlated with student outcomes and effects largely dependent on the measures employed (Pike et al. 2011).

Faculty members are asked to improve the "look and feel" of their online presence without any evidence about or knowledge of what that means beyond having more attractive landing pages or more multi-media in their online courses. Specific evidence-based determinations about the quality and efficacy of these activities are replaced with Goldilocks (too much, too little, just right)[3] estimates of the amount of time that should be allocated to get the work of design and delivery done. One thing that can be assured under these circumstances: Every time faculty members are asked to do something else to get learning and teaching "just right" more resources are required and costs increase.

Their limited knowledge about the quality of learning and teaching does not stop universities developing new and costly ways to spend money automating the status quo of pre-contextual learning and teaching. We saw this in Chap. 1 where Pre-contextual Grandbridge procured a new learning management system in the vain hope that it would transform its performance. The university engaged in an administrative re-organization (described in Chap. 5) and procured new productivity software in this chapter all designed to make the university more productive. In each example, determining the efficacy and quality of learning and teaching was not on the Pre-Contextual Grandbridge radar, yet it did not prevent the expenditure of significant resources in the hope that better learning and teaching would result. What is evident in each of the examples, is the same symmetry we have described throughout the book where the architecture of the organization drives its selection of methods and tools to create a self-fulfilling construction of what it means by learning and teaching. In this chapter, we see an extension of this symmetry to the measurement of productivity. Pre-contextual Grandbridge is measuring productivity without explaining the relationship between inputs and outputs and what that relationship means in terms of quality and efficacy.

The problem with this approach, as we have described in this and other chapters, is that productivity cannot be improved unless you can produce valid and reliable workable quality and efficacy distinctions in the variables that contribute to a productivity index. In practice, this means knowing whether the lecture worked to improve student learning, whether the high production video in the Massive Open Online Course improves completion rates, or whether the new landing page in the Learning Management System improves student engagement. Without this

---

[3]The Goldilocks phenomenon in practice can be seen as another example of connoisseurship without defining standards and workable distinctions required to determine quality.

evidence, there will be little additional insight for improving graduation rates, containing costs, or deploying faculty beyond the study of existing correlates and predictors (e.g., student entry characteristics), or setting best guess Goldilocks thresholds for the cost associated with the learning and teaching work of individual universities. This is hardly the precision and performance intended or expected when measuring the productivity of higher education.

By way of contrast, the emergent feedback approach at Grandbridge SOU described in Chap. 4 generates data all of the time about design and delivery. The efficacy and quality of a Grandbridge SOU module, course or program are not restricted to static assumptions, one-off performance studies about sample modules, or the kind of connoisseur-based evaluation of university activity described in Chap. 1. Emergent feedback means that those qualities are tested whenever a module or part there of is designed and developed, and the experience shared with students. Emergent feedback is used to evolve the understanding of the Grandbridge SOU context (its schema) and refine the quality and efficacy of the modules that represent what Grandbridge SOU is learning about itself. Productivity in this context is a dynamic and adaptive phenomenon.

Finally, the comparison between the examples illustrates a key summative takeaway from the chapter and the book overall. Example 8.1 describes the enthusiasm displayed by Pre-contextual Grandbridge's CFO about the potential future of his university as a sub-contractor in a world of unbundled university services. Unbundling follows a long history of modularization in other fields (Zenger and Hesterly 1997) and as described in the example, is unencumbered by a black-box understanding of learning and teaching. The unbundling approach is an extension of the social architecture and structure of the pre-contextual university and is fully symmetrical with an effort chain approach. In fact, it is made possible by the self-similarity in the services universities currently provide in the effort chain.

In stark contrast to the unbundling at Pre-Contextual Grandbridge, Example 8.2 describes Grandbridge SOU as it prepares to launch a bid to amplify brand identity based upon the integration of its services. The marketing team is about to launch a game-changing initiative intended to establish the university as a leader among its competitors because of the totality of what it does differently. Its approach is the antithesis of Pre-contextual Grandbridge's effort to secure its future by becoming a provider of unbundled services to others. Grandbridge SOU is asserting the educational power of its unique design in order to differentiate itself from all competitors. Instead of unbundling, Grandbridge SOU is developing the kind of mature learning and teaching architecture required to make a difference for students. Its focus is on amplifying connections in its model by leveraging the integration and connectedness in its courseware and approach to showcase its impact on student outcomes. In doing so, Grandbridge SOU is addressing the lack of differentiation in higher education brands and the risk of disintermediation described by Dvorak and Busteed (2015) and discussed in Chap. 1.

Arguably, the Grandbridge SOU approach could be defined as "content" in an unbundled world and it is fully conceivable that at some point in the future it may be possible to unbundle its quality content and sell it to other providers. However, for this to happen all universities would have to cross the same border as Grandbridge SOU to professionally controlled practice in order to possess the technological maturity to deliver content that reflects a full and complete understanding of their learning contexts. Unbundling, under those conditions would reflect the shared understanding of context. As such, we are not suggesting that unbundling is completely incompatible with the SOU approach and the messages about productivity described in this chapter. The key point for consideration is about whether universities as they embrace the concept of modularization are cognizant of the huge missed step in doing what needs to be done to first to develop a mature learning and teaching technology as described in this and preceding chapters.

The takeaway message presented in the juxtaposition of Examples 8.1 and 8.2 is that better education can make for better business. By understanding what works in a productivity model in terms of efficacy and quality, resources can be pointed specifically at those things likely to improve performance while costs can be reduced in areas that are less productive in a true learning and teaching sense. For example, when you know what quality is and what learning and teaching means it takes much less time to build the professional capacity to design and deliver learning and teaching well. When a university knows what learning and teaching means in terms of transactions, systems, methods and tools, productivity improvement can be pursued in the same ways gains are made in other fields by seeking refinement in technologies, work process etc. This is profoundly different from counting inputs and outputs irrespective of the granularity of the counting process. When the former happens, the high road educational change and improvement agenda (almost always associated with new needs and more expenditure) can translate to a successful dollars and cents story for universities. Until then the overwhelming body of evidence suggests that the cost of higher education will only move in one direction as productivity remains an ambiguous and ill-defined construct because learning and teaching is so poorly understood at scale.

Takeaways

In this chapter we described productivity and its role in the SOU. The key takeaway propositions from this chapter are:

- The unsustainable escalation in the cost in higher education is driven by a failure to understand what learning and teaching quality and efficacy mean at scale in universities;
- Pre-contextual indices of learning and teaching productivity are not informed by an understanding of quality and efficacy derived from an instantiation of the learning and teaching context;
- A successful learning and teaching productivity index requires a known demonstrable relationship between inputs and outputs;
- The understanding of the relationship between the inputs and outputs in pre-contextual productivity indices is symmetrical for the poor understanding of the relationship between learning and teaching in pre-contextual universities.

- The problem of transposition occurs when higher education adopts the methods and tools from more mature fields without due consideration of the assumptions on which those methods and tool are based;
- Learning and teaching productivity can be improved when a university produces valid and reliable workable quality and efficacy distinctions in the variables that contribute to a productivity index;
- The technology (in the broadest sense) of learning and teaching in universities lacks the maturity required for the successful modularization and unbundling of university services;
- High quality and efficacious learning and teaching is not synonymous with added cost. It can be good for the business of a university.

References

ACT (2009). *ACT report: Voluntary system of accountability learning gains methodology*. Retrieved from https://cp-files.s3.amazonaws.com/22/ACTReport_LearningGainsMethodology.pdf.

Badkar, M. (2014, May 31). Here are some horrific projections for anybody who expects to pay for college some day. *BusinessInsider*. Retrieved from http://www.businessinsider.com.au/cost-of-college-in-the-future-is-scary-2014-5.

Barrett, J. (2015, June 19). What's the value of a college education? It depends. Retrieved from http://www.cnbc.com/2015/06/19/is-a-college-degree-overvalued.html.

Bowker, G., & Star, S. (2000). *Classification and its consequences*. Cambridge, Massachusetts: The MIT Press.

Businessdictionary. (2016). Productivity. Retrieved from http://www.businessdictionary.com/definition/productivity.html.

Conway, M. E. (1968). How do Committees Invent. *Datamation, 14*(5), 28–31.

Dvorak, N., & Busteed, B. (2015, August 11). It's hard to differentiate one higher-ed brand from another. *Gallup Business Journal,* Retrieved from http://www.gallup.com/businessjournal/184538/hard-differentiate-one-higher-brand.aspx.

Drengenberg, N., & Bain, A. (2016). If all you have is a hammer, everything begins to look like a nail—how wicked is the problem of measuring productivity in higher education? *Higher Education Research & Development*. doi:10.1080/07294360.2016.1208640.

Horn, M. (2014, July, 10). *Unbundling and re-bundling in higher education*. Retrieved from http://www.forbes.com/sites/michaelhorn/2014/07/10/unbundling-and-re-bundling-in-higher-education/#2b41d28879c2.

Liu, O. (2008). *Measuring learning outcomes in higher education using the measure of academic proficiency and progress (MAPP)*. Retrieved from https://cp-files.s3.amazonaws.com/23/RR-0847MeasuringLearningOutcomesUsingMAPP.pdf.

Massy, W. (2016). *Reengineering the university. How to be mission centered, market smart and margin conscious*. Baltimore: Johns Hopkins University Press.

Pascarella, E., & Terenzini, P. (2005). *How college affects students (Vol. 2): A third decade of research*. San Francisco: Jossey-Bass.

Pike, G., Kuh, G., McCormick, A., Ethington, C., & Smart, J. (2011). If and when money natters: The relationships among educational expenditures, student engagement and students' learning outcomes. *Research in Higher Education, 52*(1), 81–106. doi:10.1007/s11162-010-9183-2.

Staton, M. (2012). *Disaggregating the components of a college degree*. American Enterprise Institute. Retrieved from http://www.aei.org/wp-content/uploads/2012/08/-disaggregating-the-components-of-a-college-degree_184521175818.pdf.

Sullivan, T., Mackie, C., Massy, W., & Sinha, E. (2012). *Improving measurement of productivity in higher education. Panel on measuring higher education productivity: Conceptual framework and data needs.* Committee on National Statistics and Board on Testing and Assessment, Division of Behavioral and Social Sciences and Education. Washington, DC: The National Academies Press.

The Economist (2012). *Not what it used to be.* Retrieved from http://www.economist.com/news/united-states/21567373-american-universities-represent-declining-value-money-their-students-not-what-it.

USDOE (National Center for Education Statistics). (2014). *Digest of education statistics, 2014.* Retrieved from https://nces.ed.gov/programs/digest/

Zenger, T., & Hesterly, W. (1997). The disaggregation of corporations—selective intervention, high-powered incentives, and molecular units. *Organization Science, 8*(3), 209–222.

Chapter 9
Moving Forward with the Self-organizing University

9.1 Introduction and Overview

The goal of the preceding chapters has been to make a case for a different kind of university, one that can self-organize to act upon the quality of learning and teaching at scale. Each chapter has engaged in strong critique of the status quo of learning and teaching in pre-contextual universities followed by the description of an alternative approach—the self-organizing university (SOU). In this chapter, we describe the forces that drive change in higher education and those that sustain the status quo. We then take up the shifts in thought and action required to create a self-organizing university. We introduce nine shifts required to build a new schema that migrates the conversation about change in higher education from the pre-contextual to the SOU.

9.2 Change Forces?

Anyone who has attended a higher education conference recently will without doubt have experienced at least one "fear and consumption" presentation. These presentations usually begin with the presenters amplifying the profound change threatening the current condition of the field including an array of statistics about out of control costs, highly competitive nontraditional providers, disintermediating technologies, change in consumption patterns, and unique generational shifts in student characteristics. These forces for change are presented as a call to action in the face of profound disruption and the vulnerability of the institutions the audience represent.

Audience members including educational leaders pressed by time-dependent institutional key performance indicators, tenured faculty members, and middle managers desperate to keep up with whatever is going on in the field await patiently

A. Bain and L. Zundans-Fraser, *The Self-organizing University*,
DOI 10.1007/978-981-10-4917-0_9

for the solution. Invariably the fear evoking part of the presentation sets up a rationale for the 'new new' thing.[1] This is the innovation that will change everything and assuredly transform the field—flipped university classrooms, new software platforms for massive online courses, or even drones for educational assessment. The basic form of these presentations has not changed at least since the advent of PowerPoint™[2] in 1990 when presentation technology made sharing both the fear and consumption parts of the deliverable even more evocative with high-quality graphics, animation, and video. Further, and as we have noted throughout the book, none of the solutions offered up in the consumption part of the presentations have turned out to be transformational nor have individual institutions produced solutions that would distinguish their learning and teaching at scale in unique ways. Despite this poor track record, proponents of the fear and consumption genre seem to reappear unperturbed by the failure of the last transformational innovation.

While the fear and consumption presentations most often offer up simplified realities and silver bullet solutions, there is little doubt that rapidly escalating costs and new technologies among other things are driving genuine disruption in the sector. We described some of those pressures as they relate to quality and productivity in Chap. 8. The key question for those concerned about learning and teaching pertains to what the disruption produces in terms of a deeper understanding of the context, efficacy, quality, and productivity of learning and teaching. As we also noted in Chap. 8, it is easy to envisage universities jumping on board with innovations like the unbundling of services, or microdegrees, and massive online open courses (MOOCS) without having to take up more fundamental change in the quality and productivity of their core activity. In fact, the escalating costs of higher education described in Chap. 8 and associated concerns about value and student capacity to pay make big change related to cost savings highly likely.

The impact of such change could be so extreme as to cause the demise of many universities. According to, Harden (2012) "In fifty years, if not much sooner, half of the roughly 4500 colleges and universities in the United States will have disappeared (p.1.)" Harden suggests this is likely because of the way technology has disintermediated any business that sells information (e.g., online trading platforms, travel sites, etc.). For example, Pre-contextual Grandbridge University first described in Chap. 2 with its limited resources, weaker reputation and extensive experience in online learning could decide to share its online capability. In an extension of example 8.1, the leadership at Pre-contextual Grandbridge may decide to outsource its online services to Pre-contextual University B. The latter, is a late starter in online education, has low online experience and capability although possesses a strong and selective enrolment. It is conceivable that Pre-contextual Grandbridge may decide its future is best secured by becoming a provider of online services to University B and others. In this scenario, the universities served by

[1]From the title of the book by Lewis (1999), *The new new thing: A Silicon Valley story.*

[2]PowerPoint [Computer Software]. (2015). Microsoft Corporation.

Pre-contextual Grandbridge's online capability broaden and strengthen their share of the market, while Pre-contextual Grandbridge could ultimately disappear as a direct service institution.

Harden's claim of the disappearing university comes true in this example as a result of market forces along with the modularization of services and unbundling. However, and most alarmingly the scenario offers no change in terms of the quality of learning and teaching and student learning outcomes. A profound change (i.e., a university goes out of business as an independent entity) occurs without any alteration to the quality of the core activity in the universities involved. The CFO at Pre-contextual Grandbridge would no doubt claim huge productivity benefits in this scenario despite the prospect that the university no longer exists as an independent higher education institution!

It is altogether possible that a host of similar or dissimilar innovations could result in the dissolution of universities and even the disintermediation of establishment faculty in those institutions that survive. In one sense, it is hard to deny that these changes are transformational. There can be nothing more transformational than going out of business! The fear and consumption presenters at the conferences can assert "we told you so." Yet as we have seen in the examples here and throughout the book student learning and teaching can remain stunningly unaltered in the process. If change in the "form or face" of higher education is inevitable, the key questions are those alluded to in the opening of this section: What kind of change are we looking for and what kind of change will improve the quality and efficacy of learning and teaching for students?

9.3 Equilibrium Forces

We see four visceral equilibrium forces that sustain the current condition of learning and teaching in higher education even if its form and face were to be radically altered as described in the previous example.

9.4 The Aptitude Mask

The first is the masking effect of selectivity and aptitude. Admissions criteria and students' pre-existing aptitude, mask the contribution higher education makes to the student learning experience. Some students will be highly successful at any kind of university and given time most can survive the experience despite huge variability in the quality of learning and teaching. This occurs because of the aptitude and experience students bring to the learning and teaching interaction. As noted in Chap. 1, student entry characteristics explain the largest amount of variability in student learning (ACT 2009; Liu 2008; Pascarella and Terenzini 2005). The masking effect of aptitude means that universities can claim a significant degree of success by simply bringing academics and students together in a loosely defined and largely idiosyncratically constructed learning and teaching experience. Universities with strong traditional reputations, built on their historical success selecting high-aptitude students use selectivity as their best "learning and teaching practice" and ongoing source of leverage in the marketplace.

To claim success under these circumstances is like a hospital admitting only healthy people and then claiming victory over disease. It would be interesting to gauge the success of universities if they worked under conditions where achieving the stated expected outcomes of a degree were somehow solely a function of the value added by the university learning and teaching experience. One of the most disturbing outcomes of a system that is driven by aptitude/selectivity over the efficacy and quality of university teaching and learning is the way students are labeled or categorized as being at risk or discrepant from accepted norms because of entry characteristics. Those characteristics are prioritized over the need to pay much more attention to the learning and teaching they experience as the driver of their success or discrepant performance.

9.5 All You Need Is a Ph.D.

A second equilibrium force is the profound resistance to or oversight of research evidence as a basis for decision-making about learning and teaching. The problem manifests at all levels of the university. For example, in Chaps. 2 and 5, we described the way the leadership at Pre-contextual Grandbridge deferred to high-level strategic plans and outside consultants to generate structural change largely uninformed by evidence about efficacious quality learning and teaching practice. As we noted, in Chaps. 4, 5, and 8, many academics do not believe that differential educational effects exist or that there is a body of professional practice associated with the field. This perspective undermines efforts to understand the learning and teaching context with widespread consequences related to determining attribution, measuring feedback, and determining quality and productivity. It also sets an immensely low and diffuse baseline for what constitutes accepted and acceptable practice.

The resistance-to-evidence force also has wider implications. For example, it diminishes the likelihood we will see technologies that represent a nexus of technological capability, evidence-based professional practice, and the individual agency described in Chap. 7. This is the interaction that has made technology a genuine transformational force in other fields where technology and their extant professional practice is combined to build remarkable new capacities. The missing professional practice component in higher education technologies most often limits the use of technology to automating existing practice. Just as embedded design means embedding key features of evidence-based practice in the organization's structures, systems, methods and tools, the absence of evidence-based practice means that those same structures, systems, methods, and tools are only distally connected at best to the things that influence student learning. This depowers a university's capacity to influence learning and teaching at scale. The appetite for genuine deep change in learning and teaching is low resulting in the kind of selective, elective, and exemplary initiatives described in Chap. 1 that cannot produce whole-of-organization effects.

9.6 The Consumer Narrative

A third equilibrium force pertains to the prevailing schema and associated narrative for learning and teaching in the broader community—the consumers of higher education. These schemas continue to reflect messianic caricatures of exceptionalism and inspiration (e.g., Dead Poets Society, Lean On Me, Freedom Writers[3]). The allegories in these stories about passion for learning and commitment are often legitimately inspirational although without intending to do so, they shape the community's perception that a great teacher simply needs to be interesting, skilled in a content area, and passionate. It is ironical that universities as institutions associated with the progress of knowledge are complicit in the perpetuation of a preprofessional paradigm in learning and teaching—their core activity. Under these circumstances, those advocating for learning and teaching change based upon professionally controlled practice are in possession of an answer to a question that is not necessarily being asked by the consumers of education or even those leading and working in universities. As we noted in Chap. 6, there can be nothing more wicked than a solution to a problem that is unanticipated, misunderstood, or ill-defined.

9.7 Impersonation

The fourth equilibrium force is an impersonation condition where the field of higher education claims "we already do that" by implementing many of the things that more established professional fields do without addressing the profound differences in the condition of the field that render most of these efforts largely ineffective at scale. We introduced numbers of the contributors to the impersonation force in Chap. 1 when discussing effort chain standards, determining quality using student evaluations, promotion frameworks, peer mediation, and the work of entities within universities to promote learning and teaching. The influence of the impersonation force reflects powerful co-dependent and co-reinforcing relationships that are highly resistant to change. For example, standards agencies reluctant to bring professional control to standards frameworks are supported by universities that lack

[3]These are titles of popular films that depict the role of teachers.

the appetite for comparable and visible professionally controlled practice (Bowker and Star 2000). It should come as no surprise that a university where faculty members and leaders believe all you need to be a good teacher is to hold a Ph.D. in your discipline (e.g., Scott and Scott 2015) would find appealing, standards that are as generic as possible in terms of visible and comparable learning and teaching expectations. Such a university would no doubt find the professionally controlled standards common to other fields, including ironically those in which those same faculty members assume membership (e.g., engineering, law, medicine, nursing, etc.) to be overly prescriptive, inhibiting of flexibility, innovation, and freedom of operation. The result of this co-dependence/co-reinforcement is standards frameworks that exert little influence on the quality of learning and teaching at scale and a perpetuation of the equilibrium condition of the field. There is a difference between clarity, professional control and prescription. Clarity is definitely a friend of quality which may explain why quality is so difficult to measure in learning and teaching in higher education (Sullivan et al. 2012).

9.8 Edge of Chaos?

What is most clear from a description of the four equilibrium forces is they will continue to exert an influence irrespective of unbundling, modularization, delivery mode (e.g., online face-to-face, etc.), staffing models, fear and consumption presentations and even given the demise of many institutions. On one hand, as we noted earlier, it is easy to imagine a disruptive square wave condition[4] where concerns about cost, capacity, and new technologies, and the demand for value, drive change that forever alters the form and face of higher education. In the evocative parlance of chaos and complexity theory, the conditions facing higher education could be seen to lure the field to the edge of chaos,[5] whereby the profound drivers of change plunge the field into a phase of profound disruption and turbulence from which a new order emerges.

However, there is little evidence to suggest those disruptive forces purportedly attracting the field to the edge of chaos will alter the equilibrium forces that sustain the current learning and teaching metaphor even if the form and face of higher education changes in profound ways. Further, there is no guarantee that what emerges in terms of new approaches will necessarily be better or more effective than that which currently exists. For example, the advent of the for-profit online university with more heterogeneous enrolment and "blue-sky" conditions for entry to the field has not produced paradigm shifting innovation in learning and teaching.

---

[4]A highly dangerous ocean condition where waves can come from all sides because of a cross-sea state.

[5]A term coined by Doyne Framer and described by Mitchell Waldrop in the 1993 book *Complexity: The emerging science at the edge of order and chaos.*

In many ways those institutions have shown how brittle the current system is when enrolment becomes more heterogeneous and the reliance on government funding is tested (e.g., Surowiecki 2015; Wong 2015). There is no guarantee that a system capable of exerting a positive influence on student outcomes at scale will emerge from the chaos as change forces play out. In fact, the history of the field would suggest there is a greater likelihood of the reappearance of existing metaphors for practice at a cheaper price point under different structural conditions (e.g., online). As such, and given the change and equilibrium forces at play in higher education what then are the shifts in thinking and action required to effect learning and teaching at scale and produce a self-organizing university?

9.9 Shift 1: Assuring Scope and Depth

The term transformational change is frequently used to describe major change in education. However, to the extent that transformation is synonymous with a comprehensive reformulation of practice and the processes that support it, few if any major educational changes meet these requirements. In Chap. 3, we indicated that change in education pivots on altering the professional lives of those involved. In that chapter we emphasized that few changes genuinely alter, the full scope of work in the area of reward, recognition, career trajectory, governance, technology, core professional practice, agency, and feedback that have been the subject of the preceding chapters. These are the changes we contend are necessary to address fully the professional lives of educators in transformational ways. Even movements like the US two billion dollar New American Schools reform effort that had a transformational mandate, for the most part functioned within the constraints of the operating paradigm of schools, their leadership approaches, funding models, and management structures (Berends et al. 2001).

As noted by Tyack and Cuban (1995) the big changes in education have been done to the field not by it through change in laws and regulation as opposed to that which the field has envisioned for itself. The outcome has been a change in the regulatory framework with limited impact on the practice of schooling at scale. We have highlighted and echoed throughout the book a similar set of circumstances in higher education where despite claims of distinctiveness and uniqueness of mission, sameness seems to be the most enduring quality of higher education institutions and where change is largely elective, selective, and exemplary. Any individual or entity embarking upon a transformational learning and teaching change process needs to first study the history of change in education and the stories of so many incomplete efforts. Any process needs to respect those involved by ensuring that a change model possesses the scope and depth required to deliver efficacious, quality learning and teaching at scale. Such change requires embedded design if it is to be transformational.

9.10 Shift 2: Migration Over Re-Engineering

We contend that a model for changing higher education at the scope and scale implied by the SOU will not be premised on re-engineering or redesigning the status quo. This is consistent with the experience of many other fields and in the invention of new products and services. Online music services did not emerge from re-engineering cassettes or CDs. The advent of modern scientific medicine was a profound and radical departure from the longstanding prevailing assumptions and practices of ancient Greek medicine. This is not to say that those innovations did not refer in some ways to what has gone before but they did not re-engineer, redesign, or reformulate the existing object of change. As posited by Buckminster Fuller, change rarely happens by fighting the status quo. A better approach is to create a new way that makes the existing reality obsolete (Fuller and Kuromiya1982).

Throughout the book we have juxtaposed Pre-contextual Grandbridge with Grandbridge SOU. This does not mean that Pre-contextual Grandbridge gets re-engineered in the SOU approach. In its earliest iteration Grandbridge SOU would emerge as a small scale alternative to the existing university that grows as a result of a migration process. It is altogether possible that the inspiration for such an initiative could come from an existing entity as described in Chap. 2. However, its progression would involve the creation of something new over any incremental improvement on the existing Pre-contextual Grandbridge.

The concept of migration offers no absolute guarantees as a change methodology for better learning and teaching. A key observation from the rise and fall of for-profit universities (mentioned earlier) or the trajectory of the MOOC movement is the way the conversation around those innovations focused on issues of cost and access over the transformation of the professional practice of learning and teaching or even the nature of learning and teaching as a driver. The design of new technology platforms for the delivery of MOOCs offer little more than the learning management systems that were their immediate antecedents. Migration does not guarantee a successful outcome. Unless an understanding of the learning and teaching context is the centerpiece of a university change opportunity, the most likely outcome of any change is an organization with new rhetoric that looks remarkably like its antecedents. Further, the bar for building a new model that makes the old one obsolete is high. Arguably, and as we noted in Chap. 8 the obsolescence of the pre-contextual university will only emerge if a new approach can produce more, deeper, faster learning. There is immense complexity in achieving and then benchmarking more, deeper, and faster learning. This is especially the case for a commodity that many sign up for and then try and get as little as possible for their investment![6] We refer here to the finding described in Chap. 8

[6]An earlier exposure to this idea came from Edward Hooper a former Director of Learning Skills at Brewster Academy, Wolfeboro New Hampshire.

where over one- third of students sought out courses with less than 40 pages of reading for an entire term (The Economist 2012).

9.11 Shift 3: Avoiding Distractions

There is a distracting quality in much of what is written in higher education about change. This includes argument about political interference in the workings of higher education (e.g., Hil 2015; Jarvis 2014) to admonitions about institutional resistance described in Chap. 3 or instrumental concerns about the implementation of change processes. These approaches can be a distraction from the more fundamental concerns related to improving learning and teaching at scale in universities. For example, the discourse about the way universities have been captured by this or that political or economic model misses the point that irrespective of the dominant overarching political context or policy paradigm, there would be little change in the performance of pre-contextual universities because of the much more significant fundamental issues described in this book. If you do not know what learning and teaching means at scale and cannot act to improve student outcomes in scalable ways, it does not really matter whether universities are run by Marxists or neo-liberals, there will be little improvement in the quality of learning and teaching. This circumstance stands in contrast to fields like medicine for example that seem to be capable of delivering patient outcomes under a range of political and policy regimes. We recognize in making this statement the existence of extensive debate about the relative merits and quality of different health systems particularly in relation to access. However, our point here is this: While it is possible to obtain a successful joint replacement in Cuba, China, Australia or the USA, you will not find a university in any of those countries that can demonstrate empirically the effects of its teaching on learning at scale.

Similarly, if you do not know what needs to change and what an alternative may look like at scale, no amount of attention to the instrumental dimensions of a change process will produce the desired outcome. Explaining the low levels of success in transformational change because of instrumental lapses or points of breakdown in change management (e.g., lack of leadership, failure to communicate, or build support) draws attention away from the primary problem. The kind of border crossing required to create a self-organizing university means sorting symptom from cause to first fully understand and articulate what a self-organizing university is/means. Good change process is important although no substitute for, nor should it be confused with an inability to articulate what a change is about. A battalion of change managers cannot substitute for a lack of understanding of that which needs to be changed, any more than a political or economic philosophy can be blamed for the pre-contextual circumstances of universities.

9.12 Shift 4. A Commitment to Understanding Context

Central to all professional fields is a commitment to the value, efficacy, and quality of their core practice. Doctors, lawyers, and engineers among others are acting upon evidence-based practice that has an attributable effect on outcomes. The evidence base is dynamic and subject to development and revision. However imperfect it becomes the basis for practice at scale at any point in time. The core issue of an effort chain approach including weak standards, resistance to professional practice, underpowered technologies, and poor evaluation methodologies is the failure to understand and act upon the learning and teaching context cycle described in Chap. 1. That cycle is the basis for defining quality, producing workable distinctions in practice and determining quality and productivity. As described throughout the book, committing to building an understanding of context is an immense challenge. This is especially the case when the field has established a degree of comfort and credibility claiming quality and expertise without paying attention to context.

9.13 Shift 5: Rethinking Feedback

A self-organizing university can demonstrate its understanding of the learning and teaching context through the systems, methods, and tools it uses to generate emergent feedback. One is only possible with the existence of the other. In our account of emergent feedback, we described the importance of feedback at its most fundamental level. This is feedback that emerges from the interaction between learner and teacher. Every piece of data in an emergent feedback approach is imbued with the intelligence and agency conveyed by the professional context from which it was derived and the student, teacher, program, course, or learning experience to which it is applied. From an emergent feedback perspective this kind of intelligent data is powerful because the context is an intrinsic component of the agency of the people who work within it. The data is part of the context, and the context is part of the people (Bain and Drengenberg 2016, p. 69).

The field of learning and teaching in higher education is caught in a feedback trap, whereby measurement is seen as a solution to its failure to define what it is trying to measure. The inability to bring clarity to the object, in this case learning and teaching means that measures are destined to be distal and inadequate, and as we described in the example of learning analytics, a case of measuring what we have over what we need. We described numerous examples of this problem throughout the book including the use of student evaluations and peer moderation of assessment at Pre-contextual Grandbridge. In each case, Pre-contextual Grandbridge desperately needed high-quality measures of learning and teaching to support bold strategic planning goals. Yet, the inability to understand what was being measured paralyzed the attempts to show the effects of the university's efforts and left it with few alternatives when things did not work out well. In the examples, we described the measurement issues as a symptom of a failure to understand the learning and teaching context. No amount of effort to build better measures can solve this problem.

9.14 Shift 6: Recognizing Small Really Is a Big Deal

Synonymous with our claims about the importance of understanding the learning context and emergent feedback is an equally important idea: Context must be understood at the most basic or fundamental level in interactions among learners and teachers. We made this point in Chap. 5 when describing the challenge of solving a problem with assessment rubrics. Further, the agency of the organization is no more than a network of understanding and action related to those interactions.

All of Pre-contextual Grandbridge's problems emanate from its failure to understand learning and teaching at a small scale. Its investment in the effort chain; its inability to reconcile strategy and execution; its blunt organizational design, and weak technological capability each reflect the inability to recognize and act upon the importance of learning and teaching at a transactional level. By way of contrast, the high value placed on learning and teaching capacity, the role of emergent feedback, networked organizational design, and the edge technologies (described in Chap. 7) all emerge from Grandbridge SOU's understanding of what learning and teaching mean when teachers and students interact. As we noted throughout the book, the idea of scale really means adjusting the field of view where leadership is mostly about acting on feedback that emerges from what happens when teachers teach and students learn. This emphasis on the small scale may seem inconsistent with the focus on scaled up whole-of-organization impact that recurs throughout the book. The point here is this: The field of view for learning and teaching can only be adjusted to depict the whole organization when the context is understood at its most basic transactional level.

9.15 Shift 7: Rethinking the Role of Technology

In Chapters 6 and 7 we offered up a profoundly different conceptualization of technology to the standard "tool use" characterization common at all levels in the field of education. We used the limitations of learning analytics and learning management systems to illustrate what happens when technology lacks agency derived from an understanding of the learning and teaching context. We described technology as an actant (Latour 1996) , a source of dynamic agency in an SOU. The agency of technology resides in the interaction among context, technology, and user where the technology influences human behavior and capability interactively (Verbeek 2006). To understand the role of technology in an SOU is to first understand the learning and teaching context and the way it can be leveraged by technologies. We described this idea in practice in Example 7.2 showing how a design tool can engender better learning and teaching practice as Jane went about designing a cooperative learning experience. Both Jane and the design tool possessed agency that built capacity with Grandbridge SOU's learning and teaching approach.

As noted earlier, the understanding of context and professionally controlled practice (Bowker and Starr 2000) is the key driver in those fields that have been transformed by technological innovation. In those fields, an understanding of efficacious practice at scale is leveraged by technological innovation. Rethinking the role of technology means giving serious consideration to the ways technology currently serves learning and teaching and whether the current circumstance is more about what we have over what we need. Genuine consideration of what we need requires understanding the learning and teaching context to drive a shift in the kind of technologies we develop for learners and teachers.

9.16 Shift 8. Assessing the Ever Present Risk of Functional Stupidity

"Functional stupidity is organizationally supported lack of reflexivity, substantive reasoning, and justification. It entails a refusal to use intellectual resources outside a narrow and 'safe' terrain" (Alvesson and Spicer 2012 p. 1196). We saw this phenomenon at Pre-contextual Grandbridge in Chap. 6 when the university reiterated a failed strategy in the face of poor initial strategic planning outcomes. Instead of considering how it could pool its intellectual resources to do things differently it adjusted its strategic plan by doubling down on initiatives that had failed to produce the intended outcomes in the first instance. Functional stupidity has benefits—it is functional. It can provide a sense of consistency of action and process that enable organizations to function smoothly and promote career interests without the disequilibrium produced by reflexive concerns and divergent courses of action (Alvesson and Spicer 2012).

Possibly the best example of a lack of reflexivity, substantive reasoning and justification in the conduct of higher education pertains to the measurement of productivity. In Chap. 8 we described the way those measuring productivity perseverate with deep analysis of how academics spend their time without establishing whether that expenditure exerts a differential effect on the quality of what those academics do and its effect on student outcomes. The precedent for measuring productivity successfully in other fields along with the pressing demand to better understand the escalating costs of higher education causes economists to apply productivity models to education without the workable distinctions in practice required to determine quality.

Throughout the book we described many other examples including using connoisseurship and benchmarking to judge program quality without adequate standards; learning analytics that do not measure learning (or teaching); and separating strategy from execution in strategic planning. In each case the members of the Pre-contextual Grandbridge community engaged in activity that could be seen to be both functional and quite stupid. Pre-contextual universities are at heightened risk for functional stupidity because of their low pre-existing knowledge base about learning and teaching. This inhibits schema development at the organizational level and the cognitive capability to act at organizational scale. Under these conditions, there is a heightened likelihood of diffusion of intellectual resources related to learning and teaching, and deference to the existing organizational order as a proxy for a deeper understanding of better practice.

Awareness that an organization can function purposefully in ways that are stupid is an important pivot for reflecting on the work of the pre-contextual university. Assessing the extent to which the circumstances of the pre-contextual university are synonymous with functionally stupid approaches to organizational decision-making can set a baseline for thought and action necessary to make better use of cognitive capabilities and the kind of knowledge building necessary for the SOU.

9.17 Shift 9: Rethinking Exceptionalism

Our final shift in thought and action pertains to the exceptionalism perspective that underpins much of the resistance to comparable and visible professionally controlled practice (Bowker and Star 2000) in higher education. Exceptionalism in this context is broadly defined as the belief that education is different from other fields in the way it takes up practice, professionalism, standards, etc. An exceptionalist focus enables a perpetuation of the status quo where universities claim learning and teaching success by touting teaching award winners and other perceived distal correlates of quality (e.g., grant recipients, facilities, etc.). An exceptionalist argument would suggest that the reason for the circumstances we have described throughout the book as pre-contextual occur because higher education is different and special. This kind of exceptionalism frequently occurs in fields that lack efficacious practice at scale and where the effects of that practice are unresolved. It is also characterized by a retreat to the relationship between professional and client/patient/student or placing the affective aspects of practice in contestation with evidence-based methodologies to mitigate or justify the lack of professionally controlled practice at scale (Bain and Drengenberg 2016).

We have been disabused of this view by studying the arguments in support of exceptionalism in the ascent of other fields including engineering, medicine and nursing (e.g., Kuhn 1996; Loudon 1997; McCloskey and Bulechek 1994). We contend that those fields at particular points in time engaged in the same kind of contestation and retreat to relationship that is common in the push-back around change in higher education. Further, the framing of the concerns in higher education do not offer up circumstances that are any more or less compelling. For those firmly entrenched in the exceptionalist camp, our intent has been to make clear that it is simply not viable to conduct learning and teaching in ways that assume universities possess an understanding of their learning and teaching context when this is so clearly not the case.

Final Thoughts

Our main purpose in writing this book has been to reframe the conversation about what a university should contribute to the learning of students. To do so we have taken up a critical analysis of the status quo and presented an alternative that is focused on a deeper understanding of what learning and teaching means and how such an understanding influences the agency of the university as a learning and teaching organization. We have shown how such agency can transform almost every aspect of the modern university from the roles of faculty to the kind of feedback that can make universities self-organizing.

In this final chapter, we have focused on the underpinning shifts in thought and action required to begin to move in a self-organizing direction. We recognize the kind of change described throughout the book first involves serious reflection and reconsideration of the current condition of learning and teaching in universities often in contexts that have a welded on commitment to the way things are. We make no apologies for the strength of our critique and reiterate that in every chapter our sometimes strident concern has been accompanied by an alternative construction of practice for each of the negative characterizations of the pre-contextual university.

While we agree with the fundamental value of "building a better mouse trap" implied by Fuller and Kuromiya (1982) we also recognize based on the ascent of other fields that transformational change is not simply a process of building a better technology and involves a range of political and affective influences. Our goal in presenting the shifts (and equilibrium forces) is to propose the change in thinking required to get to the starting line for creating an SOU.

We also wish to recognize and acknowledge the many border crossing efforts going on in universities right now that are attempting to cross over to better practice at scale. Our characterization of the status quo is not meant to dismiss or under recognize/represent or diminish those efforts; suffice to say that those of us interested in profound change have yet to produce the kind of whole-of-university alternative used as a benchmark for the arguments made throughout the book.

Finally, the terms self-organizing and self-organization are derived from complexity science and the study of complex adaptive systems. Much of the interest in complexity emerges from the apparent autogenetic nature of change and adaptation in these systems (Prigogine and Stengers 1984), especially those in the natural world and the way they adapt and self-organize without the influence of an invisible hand or top-down guidance (Waldrop 1992). This feature stands in contrast to the intentional design focus of much of what is proposed throughout this book and especially in Chap. 3 where we focus on embedded design. We contend that overwhelmingly human systems reflect acts of design and this is almost always a subtext when complexity science is applied to economies, businesses, hospitals, and other fields of human endeavor. Our goal in this book is to take up a design process the end product of which is a university capable of producing dynamic bottom-up self-organizing behavior, hence the title and the approach.

References

ACT (2009). *ACT report: Voluntary system of accountability learning gains methodology.* Retrieved from https://cp-files.s3.amazonaws.com/22/ACTReport_LearningGainsMethodology.pdf.

Alvesson, M., & Spicer, A. (2012). A stupidity-based theory of organizations. *Journal of Management Studies, 49*(7), 1194–1220. doi:10.1111/j.1467-6486.2012.01072.x.

Bain, A., & Drengenberg, N. (2016). *Transforming the measurement of learning and teaching in higher education.* New York: Routledge.

Berends, M., Nataraj Kirby, S. N., Naftel, S., & McKelvey, C. (2001). *Implementation and performance in New American Schools: Three Years into Scale-Up.* Santa Monica, CA: Rand.

Bowker, G., & Star, S. (2000). *Classification and its consequences.* Cambridge, Massachusetts: The MIT Press.

Fuller, B., & Kuromiya, K. (1982). *Critical path.* New York: St Martins Griffin.

Harden, N. (2012, December 11). The end of the university as we know it. Retrieved from http://www.the-american-interest.com/2012/12/11/the-end-of-the-university-as-we-know-it/.

Hil, R. (2015). *Selling students short.* Crows Nest, NSW: Allen & Unwin.

Jarvis, D. (2014). Regulating higher education: Quality assurance and neo-liberal managerialism in higher education: A critical introduction. *Policy and Society, 33,* 155–166.

Kuhn, T. S. (1996). *The structure of scientific revolutions.* Chicago: IL, University of Chicago Press.

Latour, B. (1996). On actor-network theory. A few clarifications plus more than a few complications. *Soziale Welt, 47,* 369–381.

Lewis, M. (1999). *The new new thing: A Silicon Valley story.* New York: W.W. Norton.

Liu, O. (2008). *Measuring learning outcomes in higher education using the measure of academic proficiency and progress (MAPP).* Retrieved from https://cp-files.s3.amazonaws.com/23/RR-0847MeasuringLearningOutcomesUsingMAPP.pdf.

Loudon, I. (1997). *Western Medicine: An illustrated history.* New York: Oxford University Press.

McCloskey, G. C., & Bulechek, G. M. (1994). Standardizing the Language for Nursing Treatments: An Overview of the Issues. *Nursing Outlook, 42*(2), 56–63.

Pascarella, E., & Terenzini, P. (2005). *How college affects students (Vol. 2): A third decade of research.* San Francisco: Jossey-Bass.

PowerPoint [Computer Software]. (2015). Microsoft Corporation.

Prigogine, I., & Stengers, I. (1984). *Order Out of Chaos: Man's New Dialogue with Nature.* New York: Bantam Books.

Scott, D., & Scott, S. (2015). Leadership for quality university teaching: How bottom-up academic insights can inform top-down leadership. *Educational Management Administration & Leadership, 44*(3), 511–531. doi:10.1177/1741143214549970.

Sullivan, T., Mackie, C., Massy, W., & Sinha, E. (2012). *Improving measurement of productivity in higher education. Panel on measuring higher education productivity: Conceptual framework and data needs.* Committee on National Statistics and Board on Testing and Assessment, Division of Behavioral and Social Sciences and Education. Washington, DC: The National Academies Press.

Surowiecki, J. (2015, November 2). The rise and fall of for-profit schools. Retrieved from http://www.newyorker.com/magazine/2015/11/02/the-rise-and-fall-of-for-profit-schools.

The Economist (2012). Not what it used to be. Retrieved from http://www.economist.com/news/united-states/21567373-american-universities-represent-declining-value-money-their-students-not-what-it.

Tyack, D., & Cuban, L. (1995). *Tinkering toward utopia: A century of public school reform*. Cambridge, MA: Harvard University Press.

Verbeek, P. P. (2006). 'Materializing Morality – design ethics and technological mediation. Retrieved from https://www.utwente.nl/bms/wijsb/organization/verbeek/materializing-morality.pdf.

Waldrop, M. M. (1992). *Complexity: The emerging science at the edge of order and chaos*. New York: Touchstone, Simon and Schuster.

Wong, A. (2015, February, 23). The downfall of for-profit colleges. Retrieved from http://www.theatlantic.com/education/archive/2015/02/the-downfall-of-for-profit-colleges/385810/.

Index

A. Bain and L. Zundans-Fraser, *The Self-organizing University*,
DOI 10.1007/978-981-10-4917-0

Kolibri 360 GmbH
Ferdinand-Jühlke-Straße 7
99095 Erfurt, Deutschland
produktsicherheit@kolibri360.de